Lecture Notes in Mathematics

Edited by A. Dold, B. Eckmann and F. Takens

Subseries: Fondazione C.I.M.E., Firenze
Adviser: Roberto Conti

1451

L. Alvarez-Gaumé E. Arbarello
C. De Concini N.J. Hitchin

Global Geometry and Mathematical Physics

Lectures given at the 2nd Session of the Centro Internazionale Matematico Estivo (C.I.M.E.) held at Montecatini Terme, Italy, July 4–12, 1988

Editors: M. Francaviglia, F. Gherardelli

Springer-Verlag
Berlin Heidelberg New York London
Paris Tokyo Hong Kong Barcelona

Authors

Luis Alvarez-Gaumé
CERN, Theoretical Division
1211 Geneva 23, Switzerland

Enrico Arbarello
Dipartimento di Matematica
Università "La Sapienza"
Piazzale Aldo Moro 5
00185 Roma, Italy

Corrado De Concini
Scuola Normale Superiore
Piazza dei Cavalieri 7
56126 Pisa, Italy

Nigel J. Hitchin
Mathematical Institute
24–29 St. Giles
Oxford OX1 3LB, England

Editors

Mauro Francaviglia
Istituto di Fisica Matematica J. L. Lagrange
Via Carlo Alberto 10
10123 Torino, Italy

Francesco Gherardelli
Dipartimento di Matematica U.DINI
Viale Morgagni 67/A
50134 Firenze, Italy

Mathematics Subject Classification (1980): 14H40, 14K25, 81E10, 32G15

ISBN 3-540-53286-2 Springer-Verlag Berlin Heidelberg New York
ISBN 0-387-53286-2 Springer-Verlag New York Berlin Heidelberg

Printing and binding: Druckhaus Beltz, Hemsbach/Bergstr.
2146/3140-543210 – Printed on acid-free paper

TABLE OF CONTENTS

Seminars

FOREWORD

"Geometry and Physics" is a binomial which has become increasingly more important in the last years and especially in the last two decades. The relations between these two subjects, whose marriage was already hidden in the *"Méchanique Analitique"* of J.-Louis Lagrange (1788), were subtly envisaged as means of understanding the structure of our Universe by the genius of B. Riemann (1854) and fully developed by A. Einstein in his celebrated theory of General Relativity (1916). Since then, although with varying degrees of fortune, the relationship between these two disciplines has grown steadily and robustously, through the contribution of many great scientists (D. Hilbert, T. Levi Civita, E. Cartan, H. Weyl, just to mention a few).

Up to a few years ago the interaction between Geometry and Physics was mainly limited to the domain of Differential Geometry, which plays a fundamental role in the local formulation of the laws of classical and relativistic field theories, and always in a single direction: from the theory to the application. More recently, however, other fundamental branches of Geometry have found their way into Physics, giving an enormous impetus especially to the investigations concerning the global behaviour of physical fields and their quantum properties, the structure of gauge theories, the theory of crystalline defects, the global structure of so-called "completely integrable" dynamical systems, as well as many other domains of application. In particular, the last two decades have seen an increasingly extensive application of Differential Topology and Global Analysis to Field Theory, and an impressive renewed role of Algebraic Geometry in both String Theory and Classical Mechanics. Also the arrow of influence has been somehow reversed, and nice results in pure Mathematics have been sometimes based on ideas originated within the context of Physics (the most striking example is, perhaps, the relatively recent work on 4-manifolds based on the structure of Yang-Mills fields).

* * * *

With exactly this spirit in mind, around 1987 the idea came to one of us (MF) to organize a Summer School on "Global Geometry and Mathematical Physics", to be held under the auspices of the CIME Foundation, with the explicit purpose of offering to both pure mathematicians and theoretical physicists the occasion of an intensive period of interaction on a number of freshly developing fields of common interest.

Also profiting of fruitful conversations with many colleagues (among which A. Cassa, R. Catenacci, M. Cornalba, M. Ferraris and C. Reina), the general structure of the Course was established. It was decided to divide the Course into four modular parts, to be assigned to experienced teachers coming both from pure Geometry and from Mathematical Physics, centering around the main themes of global aspects of field theories (instantons, monopoles, vortices, anomalies, functional integration of gauge fields) and the algebro-geometric methods in Mathematical Physics (KP hierarchies, infinite dimensional Grassmannians, theory of Riemann Surfaces and its applications to String Theory). The CIME Foundation did enthusiastically accept this program, which has been later perfectioned and worked out in detail with the collaboration of the second Editor (FG).

The Course was held in Montecatini Terme, in the beautiful and quiet surroundings of "Villa La Querceta", from July 4 to July 12, 1988. (Incidentally, this was exactly 200 years after the publication of the cited treatise of Lagrange, which laid the foundations of the fruitful interaction between Geometry and Physics and initiated the modern theory of dynamical systems of Lagrangian and Hamiltonian type). It was attended by over 70 participants, belonging to both the mathematical and the physical communities, thus reaching its aim of stimulating further interaction and development. The intensive program was centered around four main Lectures, in which all the major topics were discussed. A number of Seminars was also provided (both by participants and by invited scientists) to cover complements or parts of the subjects which could not be fully exploited during the lectures, and also to discuss some further topics having relation with the Course itself.

These Lecture Notes contain the extended text of three of the four main Lectures, as well as a carefully addressed selection of the Seminars held during the Course. It has to be remarked that all the Seminars were rather interesting and valuable, although only very few of them appear in these Notes. The choice, which has been also worked out together with the four main lecturers, was dictated by a severe limitation of space, so that here only the seminars whose contents are really complementary to the Lectures appear.

* * * *

The first topic discussed in the Course was the application of global methods of Differential Topology to the domain of Field Theory. The Lectures of Nigel HITCHIN (Oxford) on *"The Geometry and Topology of Moduli Spaces"* , were mainly aimed at discussing the various occurrences of the theory of moduli spaces in physical applications. The Lectures begin with a short description of the notion of moduli spaces, to pass immediately to discuss the self-dual Yang-Mills equations and the instantons over the 4-sphere, together with the appropriate moduli space. This opens the way to a short account on Donaldson's work on invariants for four-dimensional manifolds, which are here considered also in the general setting proposed by E. Witten. An interesting section refers to the Riemannian structure of the moduli space of all instantons, which is described in full detail together with its hyper-Kähler structure. Passing to the case of a coupled scalar field, the Yang-Mills-Higgs equations are discussed, in view of the moduli space structure of their "monopole" solutions; also the hyper-Kähler metric of this space is considered. After having discussed the above 3-dimensional and 4-dimensional situations, the Lectures address the gauge-theoretic viewpoint in the 2-dimensional case of a Riemann surface. Various important aspects are touched upon (hyper-Kähler structure, stability, existence of flat connections and related Teichmüller space structures). Finally, vortices and skyrmions, together with the structure of the corresponding moduli spaces are considered.

Some seminars were more or less directly related to the subject of this Lecture. P. HORVATHY (Metz) discussed the *"Dynamic Symmetry of Monopole Scattering"* , giving a short but detailed account about the scattering of Bogomolny-Prasad-Sommerfeld monopoles for SU(2)-gauge theories, the Hamiltonian structure of Taub-Nut geodesics and some remarkable extension to O(4,2)-symmetry. (See Cordani B., Feher L.Gy. and Horvathy P.A., Phys. Lett. **201B**, 481(1988)). M. SEPPÄLA (Helsinki) considered *"Teichmüller spaces and Moduli Spaces of Klein Surfaces"* , presenting a detailed exposition on moduli spaces of real algebraic curves, via a real version of Torelli theorem and suitable Teichmüller spaces (see M.Seppäla and R. Silhol, to appear in Math. Zeitschrift). Another group of seminars addressed the modern investigations of the "supersymmetric extension" of the motion of moduli spaces; these shall be reviewed later in this Introduction, owing to their explicit connections with other main Lectures.

Unfortunately, the Lectures of R. STORA (CERN and Annecy) about *"Differential Algebras in Field Theory"* could not be typed timely and hence could not be collected here. We profit however of a short description of their contents which has been written by the Author: -- The influence of Geometry on Quantum Field Theory has been increasing over the last few years, mainly under the influence of E. Witten, within the Theoretical Physics community, and of M.F. Atiyah, within the Mathematical community. E. Witten has created a new discipline, which one may call "Physical Mathematics", as distinguished from Mathematical Physics. The latter consisted in applied mathematics to well-formulated of some relevance to Physics. The former consist in applying mostly heuristic Field Theory or Quantum Mechanics methods to mathematical problems, mostly in Geometry, often shedding a new light on known theories and pointing to new developments. The methods used in this context, which are mostly non-perturbative, fail to incorporate one of the strongest ingredients of the perturbative methods: locality. The idea was to review consequences of locality in agreement with the geometry of some interesting models based on Lagrangians: (i) locality in perturbation theory, both in Minkowski space and in a compact Riemannian or Euclidean manifold; (ii) 4-dimensional gauge theories and their anomalies: the s-operations connected with gauge fixing; (iii) the quantization of differential forms, s-operations and their relations to the Ray supertorsion; (iv) 2-dimensional conformal models: the free string as an example, the anomaly; (v) topological theories: the topological Yang-Mills theory in four dimensions as an example (in relation with the Jones polynomials). Since then Chern-Simons in 3-dimension has proved to be one of the most interesting ones mathematically, in relation to knot theory, and most tractable models of this sort, although there are still some some subtle quantization problems to be settled --.

The problem of Chern-Simons terms in their supersymmetric version was addressed by G. LANDI (SISSA, Trieste) and U. BRUZZO (Genova) in the seminar *"Geometry of Standard Constraints and Anomalous Supersymmetric Gauge Theories"* which is included in these Proceedings as an integrating part. This was largely based on another seminar by the same authors, on *"Some topics in the Theory of Supervector Bundles"* , were the structure and cohomology of supermanifolds, super vector bundles and super line bundles was addressed, with applications to the existence of connections and characteristic classes on SVB's. Intimately related with the above topics was a seminar by R. CIANCI (Genova) on *"Differential Equations on Supermanifolds"*, which cannot be included here (see Cianci R., Journ. Math. Phys. **29**, 2152 (1988)). Still in the context of "Global Anomalies", although in a rather different perspective, we also mention a beautiful seminar by L. DABROWSKI (SISSA, Trieste), on *"Berry's Phase for Mixed States"* , referring to the rising of an extra phase in the course of various physical processes (quantum optics,...).

* * * *

The Lectures by L. ALVAREZ-GAUMÉ (CERN) form, in a sense, a bridge between the previous aspect of "global geometry applied to field theory" and the algebro-geometric side mentioned earlier in this Introduction. They were in fact concerned with discussing in great detail the mathematical structures connected with (classical, quantum and supersymmetric) string theory, as well as the whole class of so-called "conformal theories". The material presented here is divided in six parts. The Lectures contain a thorough introduction to the methods of conformally invariant theories over Riemann Surfaces of arbitrary genus g , which are considered in the operatorial approach (some of whose aspects have been retaken in the course of appropriate Seminars). In the first two sections it is shown how conserved quantities relate to representations of the "Virasoro algebra" and how Feynman's rule of integration over paths in phase-space leads to integrals over the appropriate moduli space of the relevant Riemann surface which is first supposed to have topology RxS^1. The third lecture extends to cover the case of surfaces having higher genus g and an arbitrary number n of parametrized boundaries. The corresponding moduli space $P(g,n)$ is considered in detail together with quantum states in the appropriate Hilbert space. Lecture 4 addresses the case of an interacting scalar field. In lectures 5 and 6 the author finally considers the problems of generating a connection over $P(g,n)$ out of the Virasoro algebra and of constructing a physically meaningful measure on moduli spaces of Riemann Surfaces with distinguished points.

As we already mentioned above, several specialized Seminars were devoted either to cover in greater detail some of the topics touched upon in these Lectures on Strings and Conformal Fields Theories, or to establish links with the previous two Lectures, as well as with the further Lecture of E. Arbarello. A pedagogical seminar on *"Introduction to Supergravity and Superstrings"* was given by F. GIERES (Berne). An interesting new perspective on strings was addressed in the seminar *"String Field Theory as General Relativity of Loops"* by L. CASTELLANI (Torino and CERN), whereby the dynamics of bosonic and supersymmetric strings was considered in the framework of free differential algebras on group manifolds and using loop representations based on the space Diff(S^1) of diffeomorphisms of the circle (see L. Castellani, R.D'Auria and P.Fre *"Supergravity Theory: a Geometrical Perspective"* , World Sci. (Singapore, 1989)). Strictly related to this group manifold approach and also in deep connection with Stora's Lectures on topological invariant in field theories was a couple of seminars delivered by R. D'AURIA (Padova) and P.FRE (Torino), respectively on *"Superspace Constraints and Chern-Simons Cohomology in D=4 Superstring Effective Theories"* and *"Geometrical Formulation of 4-Dimensional Superstrings"* (see, e.g., P. Fre and F. Gliozzi, Phys. Lett. **B208** , 203 (1988)). S. SHNIDER (Beersheva) gave an interesting seminar on *"Supercommutative Algebra in Higher Dimensions"* , showing in particular, in the algebraic context of Konstant's theory of graded manifolds, that no superconformal algebras exist in dimension strictly greater than six. The operator formalism for string theory in genus g larger than one, which formed the core of Alvarez-Gaumè's Lectures, was discussed in greater detail in the Seminar *"Hamiltonian Formulation of String Theory and Multiloop Amplitudes in the Operator Context"* by A.R. LUGO and J. RUSSO (SISSA, Trieste), which is here included as a complement to the Lectures themselves. On parallel lines M. MATONE (SISSA, Trieste) delivered the seminar *"Conformal Field Theories, Real Weight Differentials and KdV Equation in Higher Genus"* , which is included here; the Seminar was devoted to establish a link between the operator approach of conformal field theories and the algebraic geometric aspects related with Krichever-Novikov algebra on a Riemann surface.

Two seminars addressed, on different perspectives, the important problem of coherently defining the structure of supermoduli spaces of super Riemann surfaces, thus providing mathematically well grounded basis for the discussion of amplitudes in superstring theory: "*Super Riemann Surfaces and Super Moduli Spaces* by M. ROTHSTEIN (Suny at Stony Brook), not included here (see M. Rothstein, Proc. Amer. Math. Soc. **95** , 255-259 (1985)) , and *"Supermoduli and Superstrings"* by G. FALQUI and C. REINA (SISSA, Trieste).

These last seminars bring directly into the core of the applications of "strong" algebraic-geometrical methods in Mathematical Physics, which include nowadays a wide spectrum of techniques and domains of interest. As we already said above, the aim of the Lectures by E. ARBARELLO (Rome) was exactly to make an up-to-date review on some of these relevant topics; the Lectures *"Geometrical Aspects of Kodomchev-Petviashvily Equation"* , written together with C. DE CONCINI (Rome), address in fact all the algebro-geometric machinery involved in the KP generalization of the famous KdV equation. Their first chapter reviews the fundamental concepts from the theory of Riemann surfaces and Abelian varieties (Abel-Jacobi map, Torelli theorem, etc.). Chapter two is devoted to discuss a geometrical criterion to check whether a principal polarized Abelian variety is the Jacobian of a (possibly reducible) algebraic curve. Another criterion, which is based on the so-called "trisecant formula" and which leads naturally to the KP equation, is extensively discussed in Chapter 3. This smoothly introduces to Chapter 4, were the KP equation is used to characterize the Jacobians themselves, and to Chapter 5, where the Hirota bilinear form of the KP hierarchy is discussed. The next Chapters are finally devoted to a through discussion on the infinite dimensional Grassmannian *Gr(H)* and the corresponding t-function on the inverse determinant bundle of the Grassmannian.

A natural complement to these Lectures was a beautiful seminar on *"The Geometrical Construction of W Algebras and their Quantization"* , by D.J. SMIT (Utrecht) whereby various relations between KdV equations, Yang-Baxter equations, quantum groups and bi-Hamiltonian structures for the Virasoro algebra are discussed. We finally mention the nice seminar *"The Hilbert Schmidt Grassmannian is Non-negatively Curved"* , delivered by O. PEKONEN (Palaiseau), which refers to the Kähler structure and sectional curvature of *Gr(H)* (see O. Pekonen, Man. Math. **63**, 21-27 (1989)).

Mauro Francaviglia

THE GEOMETRY AND TOPOLOGY OF MODULI SPACES

N.J. Hitchin
Mathematical Institute
24-29 St. Giles
Oxford OX1 3LB

1. What is a moduli space?

To obtain an idea of what a moduli space is, we can go back to the original example of the space of moduli of elliptic curves (or tori). Suppose we ask ourselves the question: "What is the set of all conformal structures on a 2-dimensional torus?" Immediately our mental image is of a torus of revolution:

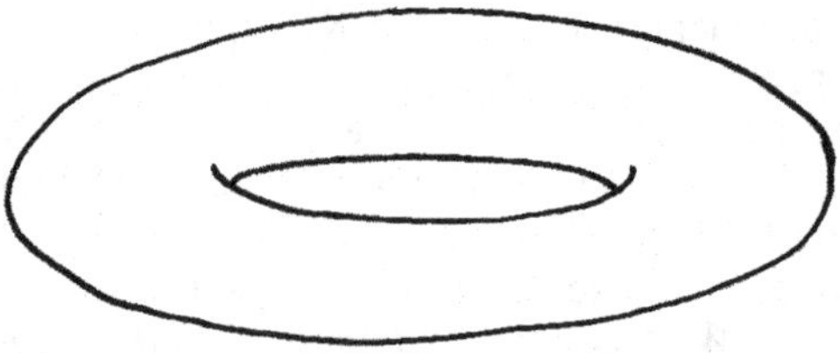

However, a moment's thought tells us that applying any diffeomorphism to the torus gives us another conformal structure:

This observation forces us to modify the original question to "What is the set of equivalence classes of conformal structures under the notion of equivalence by diffeomorphism?", or alternatively "What are the orbits of the group of diffeomorphisms acting on the space of conformal structures?"

The answer, as is well-known, is that the space of orbits is identical to the orbits of $SL(2;\mathbb{Z})$ acting on the upper half-plane. A point $\tau = x + iy$ in the upper half-plane corresponding to a conformal structure is the modulus of the elliptic curve.

The basic features we encounter here consist of an infinite-dimensional space of geometric structures (here the conformal struc-structures on a torus) and an infinite-dimensional group acting on the space (here the group of diffeomorphisms of the torus), but for which the quotient space is finite-dimensional. This quotient space has an interesting global topology (here generated by $SL(2;\mathbb{Z})$ and its subgroups) and also metric structure (the hyperbolic metric on the upper half-plane is invariant by $SL(2;\mathbb{Z})$). The degenerating behaviour of the conformal structure as we approach the boundary of the upper half-plane can also be analysed and used to produce results about the moduli space as a whole.

Somewhat remarkably, all these features are present in the moduli spaces with which these lectures are principally concerned. These are moduli spaces based on infinite-dimensional spaces of solutions to gauge-theoretic equations acted on by the infinite-dimensional group of gauge transformations. The relevant moduli spaces have on the one hand given a tool, in the hands of Donaldson, for probing the differential topology of four-manifolds and at the other extreme provided a model for the scattering behaviour of non-linear soliton-like objects. They also yield, as we shall see, a new way of approaching the old question of moduli of conformal structures on Riemann surfaces - the same moduli spaces which initiated the whole subject.

§2. The self-dual Yang-Mills equations

The first occasion in which moduli space ideas entered gauge-theory was in the study of "instantons" or self-dual solutions to the Yang-Mills equations on a compact four-manifold. A good reference for this material is the book by Freed and Uhlenbeck [FU].

We begin by taking M to be a compact, oriented, Riemannian manifold and P a principal G-bundle over M where G is a compact Lie group (for example $G = SU(2)$). We denote by $\Omega^p(M;\mathfrak{g})$ the space of exterior differential forms of degree p with values in the vector bundle $P \times_G \mathfrak{g}$ where $\mathfrak{g}$ is the Lie algebra of G. Locally, then, $\alpha \in \Omega^p(M;\mathfrak{g})$ is a Lie-algebra valued p-form.

Let us recall the basic objects associated to gauge theories. A connection on P defines a differential operator, the exterior covariant derivative,

$$d_A : \Omega^p(M;\mathfrak{g}) \to \Omega^{p+1}(M;\mathfrak{g})$$

which satisfies

$$d_A(f\alpha) = df \wedge \alpha + f d_A \alpha$$

for any C^∞ function f. The difference of two connections $d_{A_1} - d_{A_2}$ is a zero-order operator defined by $\beta \in \Omega^1(M;\mathfrak{g})$ i.e.

$$d_{A_1}\alpha - d_{A_2}\alpha = [\beta,\alpha] \quad .$$

Thus the space of all connections on P is an infinite dimensional affine space A whose group of translations is $\Omega^1(M;\mathfrak{g})$.

Associated to each connection is its curvature $F_A \in \Omega^2(M;\mathfrak{g})$ which is invariantly defined by

$$F_A = d_A^2 : \Omega^p(M;\mathfrak{g}) \to \Omega^{p+2}(M;\mathfrak{g})$$

Locally, the curvature is a 2-form with values in the Lie algebra.

A gauge transformation is a section of the bundle of groups $\underline{P} \times_{Ad} G$. It defines an automorphism of the bundle $P \times_G \mathfrak{g}$ and acts on the space of connections by $d_A \to g^{-1} d_A g$. Since d_A is a first order differential operator this is an affine action. The group of all gauge transformations G therefore acts on the space A of connections on P .

The Yang-Mills functional is the function defined on the space A of connections on P by

$$ym(A) = \int_M |F_A|^2 *1 \quad .$$

Here $*1$ denotes the volume form of the metric and the norm $|F_A|$ is obtained by using the Killing form on the Lie algebra and the metric on M. The Yang-Mills functional is invariant under gauge transformations.

The Hodge star-operator is the linear map

$$* : \Omega^p(M) \to \Omega^{4-p}(M)$$

defined by

$$*\alpha \wedge \beta = (\alpha,\beta)*1 .$$

In particular, $*$ is an automorphism of $\Omega^2(M)$ and moreover satisfies $*^2 = \text{id}$.

A self-dual form is a 2-form for which $*\alpha = \alpha$ and an anti-self-dual form is a 2-form satisfying $*\alpha = -\alpha$. An arbitrary 2-form has a decomposition

$$\alpha = \alpha_+ + \alpha_-$$

into self-dual and anti-self-dual components. The same holds for the curvature F_A, which is a Lie-algebra valued 2-form. If we write

$$F_A = F_A^+ + F_A^-$$

then the Yang-Mills functional becomes

$$\mathcal{ym}(A) = \int_M (|F_A^+|^2 + |F_A^-|^2)*1 .$$

On the other hand the Chern-Weil theorem tells us that the second Chern class, or first Pontryagin class, the basic topological invariant of a principal bundle over a 4-manifold, is given by a curvature integral. For $G = SU(2)$, we have

$$8\pi^2 k = \int_M (|F_A^+|^2 - |F_A^-|^2)*1$$

where k is an integer, called the topological charge. We therefore obtain the inequality

$$\mathcal{ym}(A) \geq 8\pi^2|k| .$$

If $k \geq 0$ and equality occurs then $F_A^- = 0$ and the connection is called self-dual. If $k \leq 0$ and equality occurs, $F_A^+ = 0$ and then the connection is called anti-self-dual. In either case the connection is a critical point of the Yang-Mills functional, the Euler-Lagrange equations for which are the Yang-Mills equations.

The moduli space of self-dual connections, or instantons, on M is the space of orbits of the group $\mathcal{G}$ of gauge transformations acting on the space of self-dual connections.

One important fact to notice about self-duality is that the Hodge star-operator on 2-forms depends only on the conformal equivalence class of the metric - scaling the metric by a positive smooth function leaves $*$ unchanged in this degree. In particular that means that the self-dual Yang-Mills equations are conformally invariant. Thus a solution to the equations on the 4-sphere S^4 with its standard round metric can be transferred to $\mathbb{R}^4$ by stereographic projection, and conversely an instanton on $\mathbb{R}^4$ with the appropriate boundary conditions can be extended over the point at infinity to lie on S^4.

§3. Instantons on S^4

Explicit knowledge about the structure of moduli spaces usually arises from explicit constructions or existence theorems. The original example of conformal structures on the torus demonstrates this - in order to obtain a point in the upper half-plane for a given torus we have to find or construct a holomorphic differential, show it is unique and that its periods determine the conformal structure up to some equivalence. Similarly, to tackle the simplest case of the moduli space of $SU(2)$-instantons with $k = 1$ on S^4, we have to construct solutions and determine parameters up to equivalence. It turns out that some essential general properties of instantons and instanton moduli spaces are also visible in this most basic situation, so we shall look in detail at 1-instantons next.

By conformal invariance we consider instantons on $\mathbb{R}^4$ and there the principal bundle P is trivial. If $SU(2)$ is regarded as the group of unit quaternions, then its Lie algebra $\mathfrak{g}$ is the space of imaginary quaternions. We also identify $\mathbb{R}^4$ with the space $\mathbb{H}$ of quaternions and write a point $x \in \mathbb{R}^4$ as

$$x = x_0 + ix_1 + jx_2 + kx_3 .$$

In this formalism we can define a connection by the covariant derivative.

$$d_A = d + A$$

where

$$A = \operatorname{Im} \frac{x d\bar{x}}{1+|x|^2} . \qquad (3.1)$$

(Of course we should check that this extends to a connection on a principal bundle over S^4.) The curvature is then

$$F_A = \frac{dx \wedge d\bar{x}}{(1+|x|^2)^2} \, . \tag{3.2}$$

(Note that

$$-\tfrac{1}{2}dx \wedge d\bar{x} = i(dx_0 \wedge dx_1 + dx_2 \wedge dx_3) + j(dx_0 \wedge dx_2 + dx_3 \wedge dx_1) + k(dx_0 \wedge dx_3 + dx_1 \wedge dx_2)$$

and so this really is a <u>self-dual</u> connection, since $*(dx_0 \wedge dx_1) = dx_2 \wedge dx_3$ etc.)

Applying translations $x \to x - a$ and dilations $x \to \lambda x$ (which are conformal transformations) to the basic instanton (3.1) we obtain a 5-parameter family of solutions whose curvature is of the form

$$F_A = \frac{\lambda^2 dx \wedge d\bar{x}}{(\lambda^2+|x-a|^2)^2} \, . \tag{3.3}$$

Consider for the moment what happens to the curvature and the density $|F_A|^2$ as λ tends to zero. The density becomes more and more concentrated around $x = a$ and in the limit becomes a delta-function:

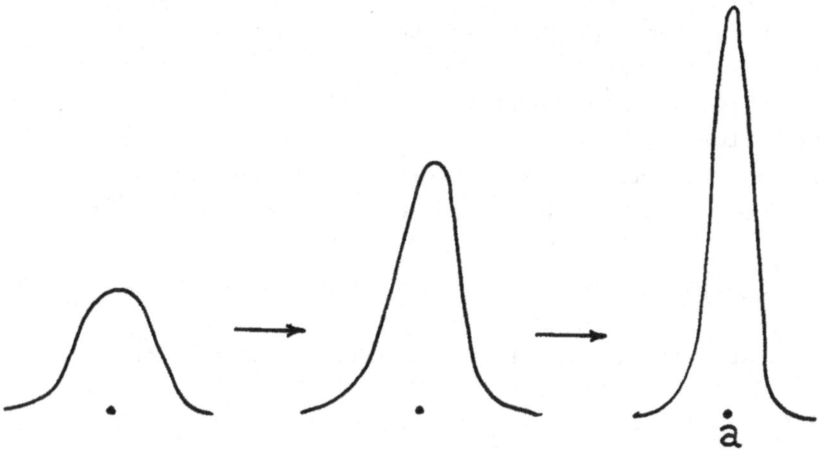

This is a particular feature of instantons in general - they can become concentrated and approximately localized.

Now one can show that every $k = 1$ instanton on S^4 is actually gauge equivalent to a member of the 5-parameter family constructed above. This is not obvious [AHS]. The proof uses, in fact, the existence and dimension of the moduli space which in this case turns out to be 5. Since the 15-dimensional group of conformal transformations of S^4 acts on this space, then we have a 10-dimensional symmetry group for each instanton and the argument proceeds from there. The analogous part of the argument for constructing the moduli of elliptic curves is the result that there is a 1-dimensional space of holomorphic differentials, which is proved by the Riemann-Roch theorem. As we shall see in the next section it is the more general Atiyah-Singer index theorem which yields the dimension 5 in our situation.

Given this fact, we see that every instanton of this class is (uniquely) gauge equivalent to one with curvature (3.3) and we can then associate to it the parameters $\lambda^2 > 0$ and a, the scale and centre of the instanton. The moduli space then appears as the upper half-space

$$M = \{(\lambda^2, a) \in \mathbb{R}^5 | \lambda^2 > 0\}$$

in $\mathbb{R}^5$. Moreover the group $SO(5,1)$ of conformal transformations of S^4 acts naturally on it.

What we have done here is to start with a particular construction, isolate parameters and then use these to construct a moduli space. There is a general construction, the so-called ADHM construction [A1], [ADHM] for an arbitrary $SU(2)$ instanton. It proceeds as follows.

We take a symmetric $k \times k$ quaternionic matrix B and a quaternionic row vector b and construct the $(k+1) \times k$ matrix

$$A(x) = \left(\frac{b}{B-xI}\right)$$

depending linearly on $x \in \mathbb{H} = \mathbb{R}^4$. We make two assumptions:

(i) $B^*B + b^*b$ is real,

(ii) $A(x)$ has maximal rank for all x.

The second condition tells us that the cokernel of $A(x)$ is a 1-dimensional quaternionic vector subspace of $\mathbb{H}^{k+1}$ for all x and hence we have a map

$$f : \mathbb{R}^4 \to \mathbb{HP}^k$$

into the quaternionic projective space. The first condition implies that the pull-back of the natural connection on the tautological quaternionic line bundle over $\mathbb{H}P^k$ is actually self-dual.

Two pieces of data (B,b) and (B',b') give gauge equivalent solutions if and only if there exist $P \in O(k)$ and a unit quaternion u such that

$$(B',b') = (P^{-1}BP, ubP) \ .$$

As an example, we can reconsider the $k = 1$ moduli space. Here the quaternion b can be reduced to a real number λ by multiplying by $u = \bar{b}/|b|$. Since $O(1) = \pm 1$ the remaining ambiguity is a sign. The 1×1 matrix $B(= a)$ is invariant and we obtain the scale λ^2 and centre a as parameters again.

Despite the apparent explicitness of this construction, it does not in fact yield much information about the moduli space for $k > 1$, largely through the non-degeneracy condition (ii). If we ask whether the moduli space is smooth, or connected, or what its cohomology groups are, then it is very difficult to use this approach. It is far better to make use of analytical techniques with a wider range of applicability and which, pushed in another direction give the deep topological results of Donaldson.

§4. Moduli spaces of instantons

Let us return now to the case where M is a compact, oriented, simply-connected, Riemannian manifold and P a principal $SU(2)$ bundle with topological charge k.

The self-duality equations for a connection A on P are the equations

$$F_A^- = 0 \ . \qquad (4.1)$$

The basic <u>analytical</u> approach to the construction of moduli spaces is to consider the linearization of these equations and then to use Banach space implicit function theorems for appropriate Sobolev spaces. To a considerable extent it is modelled on the Kuranishi construction of moduli spaces of complex structures [Ku].

To linearize the equations, note that if $F_A = d_A^2$, then an infinitesimal change $\dot{A}$ in the connection $(\dot{A} \in \Omega^1(M;\mathfrak{g}))$ gives an infinitesimal change in the curvature

$$\dot{F}_A = d_A(\dot{A}) \in \Omega^2(M;g) .$$

Thus to be tangential to the solution space of (4.1) we require

$$d_A^-(\dot{A}) = 0$$

where $d_A^-(B)$ is the anti-self-dual component of the Lie-algebra valued 2-form $d_A(B)$.

Now an infinitesimal gauge transformation $\psi \in \Omega^0(M;g)$ generates the infinitesimal change in the connection

$$\dot{A} = d_A\psi .$$

Hence the linearized moduli space, or tangent space to the moduli space, can be thought of as

$$\ker d_A^-/\mathrm{im}\, d_A \qquad (4.2)$$

where

$$\Omega^0(M;g) \overset{d_A}{\to} \Omega^1(M;G) \overset{d_A^-}{\to} \Omega^-(M;g) . \qquad (4.3)$$

The above is actually a <u>complex</u> since

$$d_A^- d_A = F_A^- = 0$$

if A is self-dual. It is also an <u>elliptic complex</u>, and so the Atiyah-Singer index theorem can be applied to give:

$$\dim H_A^0 - \dim H_A^1 + \dim H_A^2 = 8k - 3(1+b^-) \qquad (4.4)$$

where b^- is the dimension of the space of harmonic (or closed) anti-self-dual 2-forms and H^p is the p-th cohomology group of the elliptic complex. Thus:

$$H_A^0 = \ker d_A : \Omega^0(M;g) \to \Omega^1(M;g)$$

$$H_A^1 = \ker d_A^-/\mathrm{im}\, d_A$$

$$H_A^2 = \mathrm{coker}\, d_A^- .$$

The implicit function theorem referred to above, together with a collection of extra results (see [FU]) can then be used to give:

<u>Theorem</u>: <u>If</u> $H_A^0 = 0$ <u>and</u> $H_A^2 = 0$, <u>the moduli space of instantons of charge</u> k <u>is a smooth</u> $8k - 3(1+b^-)$ <u>manifold at</u> $[A]$ <u>(i.e. at the equivalence class of the instanton</u> A).

If H_A^0 is non-zero, then there exists a <u>covariant constant</u> section ψ of g. This reduces the holonomy of the connection from SU(2) to the subgroup U(1). Such reducible connections are parametrized up to gauge equivalence by their curvatures which are closed self-dual 2-forms representing integral classes in $H^2(M;\mathbb{R})$.

Understanding the non-vanishing of H_A^2 is less easy. In the case of S^4 a vanishing theorem based on the positivity of the scalar curvature and the self-duality of the conformal structure makes H_A^2 vanish, but for a general 4-manifold we have no such theorem. Instead, as shown in [FU], we can perturb the metric (so long as the connection is non-trival) to make H_A^2 vanish for all self-dual connections A. Thus a generic metric gives a moduli space which is smooth except at the reducible connections.

Moduli spaces are in general <u>non-compact</u> (like the upper half-space we saw in §3) but their non-compactness is governed by Uhlenbeck's weak compactness theorem [FU]. This states the following: if A_n is a sequence of self-dual connections of charge k, then there exists a subsequence A_{n_i} and points $\{x_1,\ldots,x_\ell\} \in M$ $(\ell \le k)$ such that A_{n_i} is gauge equivalent to a sequence B_{n_i} of connections which converge outside $\{x_1,\ldots,x_\ell\}$ to a self-dual connection of charge $\le k$.

This theorem is saying that when the instantons are concentrating around the points $\{x_1,\ldots,x_\ell\}$ (and hence failing to converge) a gauge-equivalent subsequence is nevertheless converging outside those points to an instanton of lower charge. In many applications this substitute for compactness is enough to give finiteness theorems and numerical differential topological invariants.

The original application of this information by Donaldson was in the special case where $b^- = 0$ and $k = 1$. Here, since every harmonic 2-form α is self-dual, we have

$$\int_M \alpha \wedge \alpha = \int_M {*\alpha} \wedge \alpha = \int_M (\alpha,\alpha)*1 > 0$$

hence an equivalent statement to $b^- = 0$ is that the intersection form on $H^2(M;\mathbb{R})$ (and hence $H^2(M;\mathbb{Z})$) is positive definite. Donaldson's theorem [FU] is that under these circumstances the intersection form can be diagonalized over the integers. This result follows from an analysis of the moduli space of $k = 1$ instantons with $G = SU(2)$, roughly as follows.

From the theorem above we see that, after perturbing the metric, the moduli space is a manifold of dimension $8 - 3 = 5$ except at the reducible connections. When $k = 1$, a reducible connection expresses the rank 2 vector bundle on M as $L \oplus L^*$ where the first Chern class $c_1(L)$ satisfies $c_1(L)^2 = 1$, and $c_1(L) \in H^2(M;\mathbb{Z})$ is self-dual. In this case the number of singular points in the moduli space is $\frac{1}{2} \# \{x \in H^2(M;\mathbb{Z}) \mid x^2 = 1\}$.

The most difficult part in Donaldson's proof is showing that a boundary can be put on the non-compact moduli space, the boundary being a copy of M. As a sequence in the moduli space tends to $x \in M$, the sequence of instantons it corresponds to concentrates, as in Uhlenbeck's theorem, around x. The argument to produce the topological result uses the explicit cobordism between M and boundaries of the singular points which the moduli space provides, requiring the essential feature of a natural orientation on the moduli space.

The picture we obtain from Donaldson's analysis is of the moduli space as a manifold with boundary and singular points at the reducible connections:

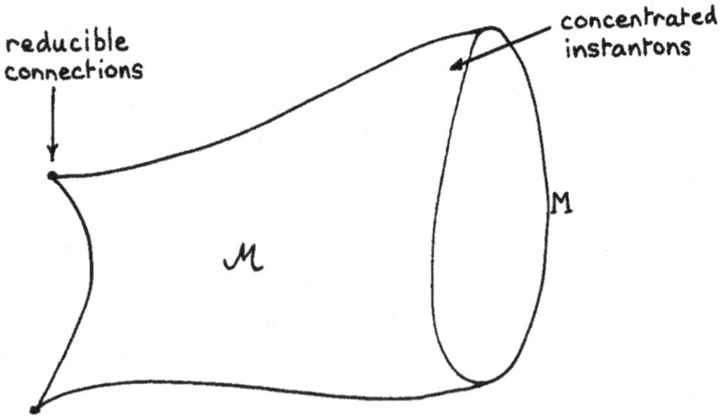

In the case of the example of $M = S^4$, which we have already considered, there are no singular points since $H^2(S^4;\mathbb{Z}) = 0$ and the moduli space is smooth without perturbing the metric, by using the vanishing theorem. If we think of the upper half-space, the moduli space of 1-instantons, as conformally equivalent to the unit ball in $\mathbb{R}^5$, then the Donaldson boundary is simply the unit sphere S^4.

§5. Donaldson invariants

Donaldson's original theorem considered 4-manifolds for which b^-, the space of anti-self-dual harmonic forms, vanishes. If $b^- > 0$, then the space of cohomology classes corresponding to harmonic self-dual forms is a proper subspace of $H^2(M;\mathbb{R})$. This can be used to our advantage: by deforming the metric we can perturb this to a subspace which contains no non-trivial integral classes so there can be no self-dual line bundles on M. This means that there are no reducible $SU(2)$ connections for $k \neq 0$ and so the appropriate moduli space is smooth.

A further regularization of the moduli space can be achieved by considering SO(3)-connections rather than SU(2)-connections. The obstruction to lifting an SO(3)-bundle to an $SU(2)$ bundle is the second Stiefel-Whitney class $w_2 \in H^2(M;\mathbb{Z}_2)$. This is actually preserved under weak limits, so when instantons become concentrated, the residual instanton produced outside the points $\{x_1,\ldots,x_m\}$ still has $w_2 \neq 0$. This means in particular that during convergence we keep well away from the trivial connection (for which w_2 is necessarily zero) and by perturbation of the metric the moduli space can be compactified by adding on lower dimensional smooth strata. One further aspect of $SO(3)$ instanton spaces is that k may now be a half integer and the dimension formula is of the form [FU]

$$\dim M = 2\ell - 3(1+b^-)$$

where ℓ is a positive integer.

These particular features of the case $b^- > 0$ led Donaldson to define new differential topological invariants of 4-manifolds.

The simplest situation is the case where $w_2 \neq 0$ and $2\ell = 3(1+b^-)$ and so, after metric perturbations, the moduli space is zero-dimensional. Of course, in general the moduli space is non-compact, but the weak compactness of Uhlenbeck's theorem here tells us that it is finite. The proof is simply to consider an infinite sequence of connections representing distinct points in the moduli space and apply Uhlenbeck's theorem. If the points do not converge in the moduli space then outside points $\{x_1,\ldots,x_m\}$ in M gauge-equivalent connections converge to an instanton of charge $\ell' < \ell$. But then since $w_2 \neq 0$ we are away from the trivial connection so $2\ell' - 3(1+b^-) < 2\ell - 3(1+b^-) = 0$ is truly the dimension of a moduli space, which is a clear contradiction. The original moduli space is thus compact and hence finite.

The existence of an orientation on the moduli space gives each point a sign and the algebraic sum of these signs is an integer which can be shown (at least for $b^- > 1$) to be independent of the generic metric and hence an invariant of the underlying differentiable structure.

A more general collection of invariants, the <u>Donaldson polynomials</u>, are based on the same idea, but are defined for <u>even-dimensional</u> moduli spaces.

We begin with a homology class in $H_2(M;\mathbb{Z})$ and a smooth embedded surface Σ representing it. Each connection A on P over M restricts to a connection over Σ and we can consider the Cauchy-Riemann operator

$$d_A'' : \Omega^0(\Sigma;K^{\frac{1}{2}} \otimes V) \to \Omega^{0,1}(\Sigma;K^{\frac{1}{2}} \otimes V)$$

defined by the natural $\bar{\partial}$-operator on the square root $K^{\frac{1}{2}}$ of the canonical bundle of Σ (with its induced conformal structure from M) and the covariant derivative d_A on the rank 2 vector bundle V associated to the $SU(2)$ principal bundle P. Specifically,

$$d_A''(s \otimes v) = \bar{\partial}s \otimes v + s \otimes (d_A v)^{0,1}$$

where $\alpha^{0,1}$ is the $(0,1)$ component of the 1-form α.

This family of operators defines (see [Q]), as A ranges over the space $\mathcal{A}$ of all connections on M, a <u>determinant bundle</u> $\mathcal{L}_\Sigma$, and since the index is zero there is also a canonical section which vanishes when $\dim \ker d_A''$ is non-zero. Restricting to the self-dual connections (and after some perturbations) the determinant section vanishes on a codimension 2 submanifold V_Σ. This is essentially a geometrical representative of the 1st Chern class of $\mathcal{L}_\Sigma$.

Now suppose that b^- is <u>odd</u>. Then the dimension $8k - 3(1+b^-)$ of the moduli space $\mathcal{M}$ of $SU(2)$ connections of charge k is an <u>even</u> integer $2d$. Taking d surfaces $\Sigma_1,\ldots,\Sigma_d$ we obtain codimension 2 submanifolds $V_{\Sigma_1},\ldots,V_{\Sigma_d}$ and hence, for a generic metric, a zero-dimensional set

$$V_{\Sigma_1} \cap \ldots \cap V_{\Sigma_d} \in \mathcal{M} .$$

If $k > \frac{1}{4}3(1+b^-)$ then a generalization of the above argument using Uhlenbeck's weak compactness shows that this set is <u>finite</u> and has an orientation and that the corresponding signed number of points n depends only on the <u>homology classes</u> $[\Sigma_i] \in H_2(M;\mathbb{Z})$. It yields a function

$$Q([\Sigma_1],\ldots,[\Sigma_d]) = n$$

on $H_2(M;\mathbb{Z})$ which is an integer-valued polynomial. There is moreover one for each instanton charge k.

These Donaldson polynomials [D5] can be used to show that certain 4-manifolds which are defined as algebraic surfaces have distinct diffeomorphism types. The key to such results is that for a Kähler manifold, anti-self-dual connections have an interpretation in terms of algebraic geometry.

Suppose, then, that the underlying 4-manifold on which we are doing our gauge theory is a Kähler manifold. Thus the Riemannian metric on M is compatible with a complex structure (M is therefore a two-dimensional complex manifold) and we have a covariant constant Kähler form ω.

The notions of self-dual and anti-self-dual 2-forms now have an alternative interpretation. With respect to the canonical orientation determined by the complex structure, a self-dual 2-form looks locally like

$$\alpha^+ = f\omega + g dz_1 \wedge dz_2 + \bar{g} d\bar{z}_1 \wedge d\bar{z}_2$$

and an anti-self-dual 2-form

$$\alpha^- = \Sigma f_{ij} dz_i \wedge d\bar{z}_j \quad \text{and} \quad \alpha^- \wedge \omega = 0 \ .$$

Thus the curvature F_A of an anti-self-dual $SU(2)$-connection on M has two properties:

(i) F_A is of type $(1,1)$

(ii) $F_A \wedge \omega = 0$.

The first property says, as a consequence of the Newlander-Nirenberg theorem, that the connection defines a holomorphic structure on the vector bundle V associated to P : local holomorphic sections are defined by $(d_A s)^{0,1} = 0$ and (i) ensures that there are enough local solutions. The second part implies, by a vanishing theorem, that the vector bundle is stable in the sense of algebraic geometry.

The converse is also true - any stable bundle on a Kähler surface has associated to it a unique solution to the anti-self-dual Yang-Mills equations. This was proved for algebraic surfaces first by Donaldson [D1], and then in more generality in arbitrary dimensions by Uhlenbeck Yau [UY] and Donaldson [D3].

The net result of this theorem is that the moduli space of instantons on an algebraic surface may sometimes be calculated by using only a knowledge of the complex structure and cohomology class of ω

(which is all one needs to define stability) and the Donaldson invariants explicitly evaluated. Some consequences of this fact include the result of Friedmann & Morgan, Okonek & Van de Ven that there exist compact 4-manifolds with infinitely many non-equivalent smooth structures.

[Note that Kähler manifolds are particularly appropriate for the application of Donaldson polynomials since the space of self-dual harmonic forms consists of the one-dimensional space of constant multiples of the Kähler form plus the complex vector space of holomorphic 2-forms and so is odd-dimensional. Thus b^- in our formula (remember the anti-self-dual connections define holomorphic bundles) is odd.]

§6. A general setting for the Donaldson invariants

A new perspective on the Donaldson invariants which were geometrically defined in the previous section has been introduced by Witten [W]. This has persuaded mathematicians to view the invariants in a rather different way and to draw analogies with finite-dimensional situations. We shall here only touch on the starting point for such a reinterpretation, which involves thinking in rather more general terms, and has been considered by Atiyah, Donaldson, Quillen and others.

We begin with the basic objects which define our gauge theory:

A - the space of connections on P

G - the group of gauge transformations

$\Omega^-(M;g)$ - the vector space of anti-self-dual g-valued 2-forms on M

$F_A^- : A \to \Omega^-(M;g)$ - the anti-self-dual component of the curvature.

In the situation where there are no reducible connections, the group of gauge transformations acts freely on A, so we can think of A as a principal bundle over the space of gauge-equivalence classes of connections. The map F_A^- is equivariant with respect to the action of G where G acts linearly on the vector space $\Omega^-(M;g)$. In finite dimensions we have the following analogous picture:

P - a principal G-bundle over $P/G = X$

V - a vector space with a linear G-action

$f : P \to V$ an equivariant map.

This data can be put in more geometric terms by defining the vector bundle $E = P \times_G V$, and then f defines a section s of E.

The basic Donaldson invariant we saw in §5 arises when the moduli space of self-dual connections is zero-dimensional. The invariant then is the algebraic sum of the signs associated to each point.

In the above viewpoint, the self-dual connections are the zeros of the function f (i.e. F_A^-), and the moduli space is the quotient $f^{-1}(0)/G$ which is the set of zeros of the section s of E. The Donaldson invariant now fits into a more familiar formalism - the algebraic sum of the zeros of a section of a vector bundle is the Euler class. The most familiar situation is the case where the vector bundle is the tangent bundle of a surface, a section is a vector field and the algebraic sum of the zeros is the index of the vector field. We know that this is given by the Euler characteristic of the surface and this is moreover also given using the Gauss-Bonnet theorem by an integral over X of a curvature. If the analogue is to be pursued, then we are required to "integrate" over the infinite dimensional space A/G (or A) a suitable form. This is such stuff as quantum field theory is made on.

One of the current lines of approach to this question is to make use of equivariant cohomology, in an infinite-dimensional context, which is closely related to the BRS cohomology of physicists. Before we describe this in the simplest situation, let us go back to the index of a vector field and the Gauss-Bonnet integrand and see more closely how they are related.

The object which mediates between the number of zeros of a section of a vector bundle E over X and a cohomology class on X is the Thom class U, which is a compactly supported cohomology class on the total space of the vector bundle E. Given a section s of E, it pulls back to a cohomology class on X itself. Considering the homologous sections ts for $t \in \mathbb{R}$, the class can be evaluated on the one hand at $t = 0$ (i.e. the zero-section) which gives the Euler class and on the other hand as $t \to \infty$ (since it has compact support) it can be evaluated in neighbourhoods of the zeros of s. Thus a good representative of the Thom class is required in order to obtain a stronger hold on the problems of infinite dimensions.

A "natural" representative could be taken to be an <u>equivariant</u> one: the cohomology of the vector bundle E is the G-equivariant cohomology of $P \times V$. To see an example of this, consider the case where $G = T$, a circle. Then the equivariant cohomology of a manifold with a circle action is given by the cohomology of the complex

$$\Omega^*_T \otimes \mathbb{R}[u]$$

where Ω^p_T denote the T-invariant p-forms and u is an indeterminate of degree 2. The differential is

$$d_T(\alpha \otimes p(u)) = d\alpha \otimes p(u) + i(\xi)\alpha \otimes up(u)$$

where ξ is the vector field generated by the action. Note that

$$d_T^2 = (di(\xi) + i(\xi)d)\alpha \otimes up(u)$$

which vanishes since α is T-invariant and so

$$L_\xi \alpha = di(\xi)\alpha + i(\xi)d\alpha = 0 .$$

In [MQ], Mathai and Quillen give a representative for the Thom class in a general version of this formalism. It is not of compact support, but instead has a <u>Gaussian</u> fall-off. If P is a circle bundle over X and $V = \mathbb{R}^2$ with the usual action by $\xi = -x_1 \frac{\partial}{\partial x_2} + x_2 \frac{\partial}{\partial x_1}$, and if θ is a connection form on P, then an equivariant Thom class representative is

$$U = \frac{1}{\pi} \cdot e^{-|x|^2}(u + \frac{1}{2}\, d\theta + (dx_1 + \theta x_2) \wedge (dx_2 - \theta x_1)) .$$

The suggestive shape of this form indicates that it has a role to play in understanding Witten's approach. However, any rigorously defined integral must necessarily take into account the compactness theorems which we have seen are necessary in order to obtain finite numbers. The exact mathematical status of the quantum field theory point of view is not yet clear.

§7. Metrics on instanton moduli spaces

Moduli spaces often inherit geometrical properties associated to the objects which they parametrize. For example, the moduli space of complex structures on a compact complex manifold is itself a complex analytic space. (Such properties are not necessarily immediately obvious, as the historical remarks in Kodaira's book on deformation theory [K] show).

In the case of instantons we need a conformal structure on the 4-manifold M to define the self-dual Yang-Mills equations, and most of the information on the topological structure of the moduli spaces involves choosing a metric within this conformal equivalence class. Suppose we choose such a metric, then there is in fact a natural induced metric - called the L^2 metric - on the moduli space.

To see how this is defined, recall the construction of the moduli space as a manifold in §4. The linearized equations gave us an isomorphism between the tangent space of the moduli space at $[A]$ and the first cohomology group of the elliptic complex

$$\Omega^0(M;g) \xrightarrow{d_A} \Omega^1(M;g) \xrightarrow{d_A^-} \Omega^-(M;g) .$$

Now Hodge theory tells us how to choose representative forms for cohomology classes. In this situation it says that there is for each cohomology class in H_A^1 a unique form $\alpha \in \Omega^1(M;g)$ such that

$$d_A\alpha = 0 \quad \text{and} \quad d_A^*\alpha = 0$$

where $d_A^* : \Omega^1(M;g) \to \Omega^0(M;g)$ is the formal adjoint of $d_A : \Omega^0(M;g) \to \Omega^1(M;g)$. Here the definition of d_A^* genuinely requires a metric on M - the conformal structure will not suffice.

Using this harmonic representative, we define an inner product on H_A^1 by integration:

$$g(\alpha,\alpha) = \int_M |\alpha|^2 *1 .$$

Since everything in sight is G-invariant this defines a metric on the moduli space (assuming it is smooth in the first place), which is the L^2-metric.

A number of authors have investigated the L^2-metric on the basic example of an instanton moduli space - the open unit ball in $\mathbb{R}^5$ which parametrizes $SU(2)$ instantons of charge 1 on S^4. They are Groissier & Parker [GrP], Doi, Matsumoto & Matumoto [DMM] and Haberman [Ha]. The resulting formula is surprisingly complicated but a number of interesting features can be read off:

(i) the metric is conformally flat - in particular it is conformally equivalent to the unit ball.

(ii) the metric extends to the boundary (the Donaldson boundary of §4) as a C^2-metric.

(iii) the boundary is isometric to S^4 and is totally geodesic.

Groissier & Parker have extended some of these results to the general situation of a compact 4-manifold with $b^- = 0$. To a certain extent it provides a metric alternative to the Donaldson "collar" theorem, since the boundary can in theory be defined in terms of geodesics in the moduli space.

A further result for instantons on $\mathbb{C}P^2$ with $k = 1$ is due to Groissier [Gr]. Here, using the explicit construction of such solutions given by Donaldson [D2] and Buchdahl [B], the L^2-metric can again be calculated. In this situation we have $b^- = 0$ but $b^+ = 1$, so there is a singular point corresponding to the reducible connection. The moduli space is a cone on $\mathbb{C}P^2$ with the parameter t along the generators of the cone. The reducible connection is at $t = 0$, the boundary at $t = 1$.

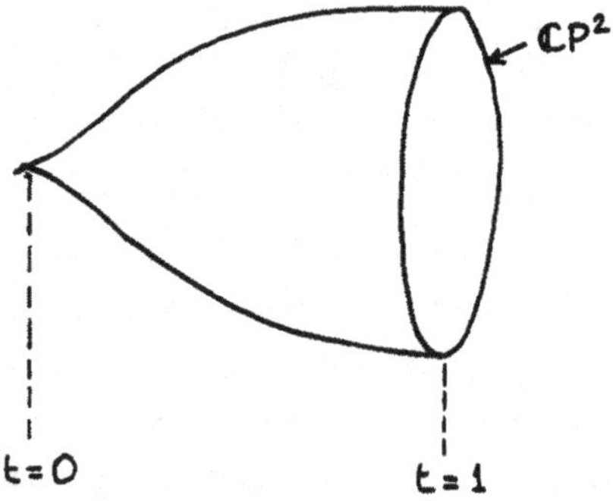

The L^2-metric has the form

$$g = 4\pi^2 (f(t)dt^2 + h(t)g_0)$$

where g_0 is the homogenous Fubini-Study metric on $\mathbb{C}P^2$ (the metric with which the instantons are defined). The functions $f(t)$ and $h(t)$ are:

$$f(t) = 2\left\{\frac{4t^{-4}-3t^{-2}}{1-t^2} + t^{-6}(4-t^2)\log(1-t^2)\right\}$$

$$h(t) = -(3-t^2)^{-1}\{t^{-2}(6-9t^2+t^4) + 6t^{-4}(1-t^2)^2\log(1-t^2)\} .$$

Here again we have a totally geodesic boundary which is isometric to $\mathbb{C}P^2$. The metric has the interesting feature that the sectional curvature is positive near the boundary but mixed sectional curvatures near the singular point are <u>negative</u>.

Clearly such calculations are limited by the need to know explicitly the solutions to the self-dual Yang-Mills equations. There is, however, a general result concerning <u>hyperkähler</u> 4-manifolds. We should recall that a hyperkähler manifold is a Riemannian manifold of dimension $4m$ whose holonomy is a subgroup of $Sp(m)$. Equivalently it has a metric g which is Kähler for three (integrable) complex structures I, J and K which satisfy

$$IJ = -JI = K$$

$$JK = -KJ = I$$

$$KI = -IK = J \, .$$

The tangent bundle of a hyperkähler manifold is thus a quaternionic vector bundle. In four-dimensions the examples are the torus T^4 with its flat metric and the $K3$ surface.

We have the following result concerning the L^2-metric for hyperkähler 4-manifolds.

<u>Theorem</u>: Let M be an (anti-self-dual) instanton moduli space for a compact hyperkähler 4-manifold. Then the L^2-metric on M is hyperkähler.

(As a remark, let us note that the three covariant constant Kähler forms trivialize the bundle of self-dual 2-forms on M and hence $b^+ = 3$. The dimension of the moduli space is therefore of the form $8k - 12$, which is clearly divisible by four).

To see why this theorem should hold we adopt a global point of view on the structure of the self-duality equations themselves. We consider, as in the case of the previous section, the space A of all connections on the appropriate principal bundle over M as an infinite dimensional manifold. Its tangent space at any point A is $\Omega^1(M;g)$ (A is an affine space) and the I, J and K of the hyperkähler structure on M make this into an infinite-dimensional quaternionic vector space. Together with the L^2 inner product induced from M, the space A is an infinite-dimensional (flat) hyperkähler manifold. The group of gauge transformations G acts on A and preserves both the L^2-metric and the operations I, J and K. This is a context where one may apply the hyperkähler moment map idea of [HKLR].

Given a hyperkähler manifold X with an action of a Lie group G preserving the three Kähler forms ω_1, ω_2, ω_3 we make the following construction. For each vector field ξ generated by G we have

$$0 = L_\xi \omega_i = d(i(\xi)\omega_i) + i(\xi)d\omega_i$$

and since ω_i is closed, we have

$$d(i(\xi)\omega_i) = 0 .$$

Hence if X is simply connected there is a function f_i such that

$$i(\xi)\omega_i = df_i .$$

Letting ξ run over the vector fields generated by the Lie algebra of G the functions f_i define a function

$$\mu_i : X \to g^*$$

which, if it is G-equivariant, is called a moment map for the symplectic forms ω_i. Putting the three symplectic forms together we obtain a hyperkähler moment map

$$\underline{\mu} : X \to g^* \otimes \mathbb{R}^3 .$$

Under these circumstances a theorem [HKLR] which generalizes the Marsden-Weinstein quotient in symplectic geometry says that the natural metric on $\underline{\mu}^{-1}(0)/G$ is hyperkähler.

What is the hyperkähler moment map for the hyperkähler manifold A with the group G acting on it? It is simply

$$\underline{\mu}(A) = (F_A \wedge \omega_1, F_A \wedge \omega_2, F_A \wedge \omega_3) \in \Omega^4(M;g) \otimes \mathbb{R}^3$$

and then

$$\underline{\mu}^{-1}(0)/G = \text{the moduli space of anti-self-dual connections.}$$

Thus the natural L^2-metric on the moduli space is hyperkähler.

This approach also works for the non-compact hyperkähler manifold $\mathbb{R}^4$ providing we use gauge transformations which are trivial at infinity. It yields for $SU(2)$ instantons of charge k an $8k$-dimensional space with hyperkähler metric. As an example, for $k = 1$, we obtain the flat metric on $\mathbb{R}^4 \times (\mathbb{R}^4\backslash 0)/\pm 1$.

Note that one universal feature of the L^2 metric is that it is incomplete: when instantons become concentrated, they do so at a finite distance from the moduli space. This may be useful from the point of view of compactifications of moduli spaces but it is unsatisfactory from the point of view of differential geometry. One other unsatisfactory aspect of the L^2-metric is that the solutions to the self-dual Yang-Mills equations are conformally invariant, whereas the L^2-metric is not - it depends heavily on the choice of metric in the conformal equivalence class.

One possible alternative to the L^2-metric is to introduce a metric which arises in statistics [Am]. This is a metric on a parameter space of probability distributions. Since the Yang-Mills density

$$|F_A|^2 * 1 = -\mathrm{Tr}(F_A \wedge * F_A)$$

is conformally invariant and gauge-invariant, and

$$\int_M |F_A|^2 * 1 = 8\pi^2 k$$

for an instanton, then

$$\rho = \frac{1}{8\pi^2 k} |F_A|^2 * 1$$

is a family of probability distributions on M, parametrized in a conformally invariant manner by the moduli space of instantons. We make the assumption (which is probably true for a generic metric) that the moduli space is a submanifold of the infinite-dimensional affine space of all smooth 4-forms of volume 1.

To define the "statistical" metric we fix a smooth measure dx on M and write

$$\rho = f(x; y_1, \ldots, y_k) dx$$

for local coordinates $y_1, \ldots, y_k$ in the parameter space, and then define

$$g = \int_M \frac{1}{f} \left(\sum_i \frac{\partial f}{\partial y_i} dy_i \right)^2 dx . \tag{7.1}$$

(In statistical terminology this is the Fisher information matrix).

The metric is clearly coordinate-invariant. It is also independent of the choice of measure dx, since if $f dx = \tilde{f} d\tilde{x}$, then $f = h(x)\tilde{f}$ and so

$$\frac{1}{f} \left(\sum_i \frac{\partial f}{\partial y_i} dy_i \right)^2 dx = \frac{1}{h\tilde{f}} \cdot h^2 \left(\sum_i \frac{\partial \tilde{f}}{\partial y_i} dy_i \right)^2 dx$$

$$= \frac{1}{\tilde{f}} \left(\sum \frac{\partial \tilde{f}}{\partial y_i} dy_i \right)^2 d\tilde{x} .$$

Note also that since

$$\int_M f dx = 1 ,$$

$$0 = \int_M \frac{\partial f}{\partial y_i} dx = \int_M f \left\{ \frac{\partial \log f}{\partial y_i} \right\} dx$$

and hence, differentiating again,

$$0 = \int_M f \frac{\partial^2}{\partial y_i \partial y_j} (\log f) dx + \int_M \left(\frac{\partial f}{\partial y_j} \right) \left(\frac{\partial f}{\partial y_i} \right) \cdot \frac{dx}{f}$$

so an alternative (and sometimes more accessible) formula is

$$g_{ij} = - \int_M \frac{\partial^2}{\partial y_i \partial y_j} (\log f) \rho .$$

Perhaps the most convincing example that justifies this definition is to take the Gaussian distributions on the line:

$$\rho(x;y,\sigma) = \frac{1}{\sqrt{2\pi}\sigma} \cdot e^{-(x-y)^2/2\sigma^2} dx$$

where y is the mean and σ^2 the standard deviation. Putting $z = \sigma/\sqrt{2}$, the metric defined by (7.1) is the hyperbolic metric $(dy^2+dz^2)/z^2$ on the upper half-plane $\{(y,z) | z > 0\}$. Note that as the standard deviation gets smaller, the distribution becomes more and more concentrated and in the limit $\sigma = 0$ we obtain as a boundary of the parameter space the real line on which the distributions are defined. This is exactly the phenomenon for the $k = 1$ moduli space described in §3 and §4. In fact, if we apply the statistical metric to the case of the moduli space of $k = 1$ instantons on S^4, then because we have built in conformal invariance, we obtain a metric on the unit ball in $\mathbb{R}^5$ which is invariant under the group of conformal transformations $SO(5,1)$. There is now no choice - it is forced to be a multiple of the constant negative curvature metric. In particular, this conformally invariant metric is complete.

Which metric one uses, obviously depends on what one wants to use it for. The statistical approach incorporates all the symmetries of the original problem and might therefore be regarded as more natural, on the other hand the L^2-metric has a particularly elegant form in the hyperkähler case.

§8. The Yang-Mills-Higgs equations

The instantons with which we have been dealing so far are all based on solutions of the Yang-Mills equations, which in turn were derived as the variational equations for the Yang-Mills functional

$$\int_M |F_A|^2 * 1 .$$

Another functional, the Yang-Mills-Higgs functional also arises in physics, partly as a mechanism for introducing mass into gauge theories, and also yields interesting moduli spaces.

In Yang-Mills-Higgs theory we have not just a connection A on a principal bundle P as our basic object, but also a Higgs field ϕ, which is a section of some vector bundle associated to P. In the general situation there is also a potential function $V(\phi)$. The Yang-Mills-Higgs functional is

$$\mathcal{YMH}(A,\phi) = \int_M |F_A|^2 + |d_A\phi|^2 + V(\phi) .$$

A common form for the potential V is

$$V(\phi) = \lambda(1-|\phi|^2)^2 . \tag{8.1}$$

The particular class of solutions to the Yang-Mills-Higgs equations we shall consider are called magnetic monopoles. These are static solutions on $\mathbb{R}^3$ in the Prasad-Sommerfield limit, which means we ignore the potential term, but keep the asymptotic condition $|\phi| \to 1$ which its presence would have forced on us.

We take then a principal $SU(2)$ bundle P on $\mathbb{R}^3$ (which is of course trivial) and a connection A on P together with a section ϕ of the adjoint bundle, (i.e. $\phi \in \Omega^0(\mathbb{R}^3;\mathfrak{g})$) and we look for the critical points of the functional of (A,ϕ) given by

$$\int_{\mathbb{R}^3} |F_A|^2 + |d_A\phi|^2 . \tag{8.2}$$

Although the bundle P is trivial, the asymptotic behaviour of ϕ defines a topological invariant. We note that since $|\phi| \to 1$ as $r \to \infty$, then in some gauge,

$$\phi \to \begin{pmatrix} i & 0 \\ 0 & -i \end{pmatrix}$$

and then the $i|\phi|$-eigenspace of ϕ defines a complex line bundle on $\mathbb{R}^3\backslash B_R$ where B_R is a ball of large radius. Since this retracts onto S^2, the bundle has a first Chern class k, which we call the magnetic charge.

Just as in the case of instantons, the magnetic charge provides a lower bound for the functional (8.2), obtained by integration by parts. First we write

$$\int_{B_R} |F_A|^2 + |d_A\phi|^2 = \int_{B_R} |F_A - *d_A\phi|^2 + 2(*d_A\phi, F_A) \tag{8.3}$$

and then consider

$$\begin{aligned} d(\phi, F_A) &= (d_A\phi,\ F_A) - (\phi, d_A F) \\ &= (d_A\phi, F_A) \quad \text{(using the Bianchi identity)} \\ &= *(*d_A\phi, F_A)\ . \end{aligned}$$

Thus using Stoke's theorem

$$\int_{B_R} (*d_A\phi, F_A) = \int_{S_R} (\phi, F_A)$$

and this is asymptotically expressed in terms of the integral of the curvature of the eigenspace bundle of ϕ over a 2-sphere and hence gives the first Chern class. We therefore have from (8.3)

$$\int_{\mathbb{R}^3} |F_A|^2 + |d_A\phi|^2 = \int_{\mathbb{R}^3} |F_A - *d_A\phi|^2 + 8\pi k \geq 8\pi k$$

if $k \geq 0$.

The absolute minimum is obtained when (A,ϕ) satisfy the Bogomolny equations

$$F_A = *d_A\phi\ . \tag{8.4}$$

A solution to these equations which satisfies the appropriate boundary conditions for the above derivation to be made rigorous we call a magnetic monopole of charge k. A good reference to the moduli space question and a guide to the literature is the book by Atiyah and the author [AH]. The analytical properties of monopoles are treated in Jaffe and Taubes' book [JT].

The basic monopole of charge 1 is given explicitly by the formula

$$\left.\begin{aligned} A_i^a &= \left(\frac{r}{\sinh r} - 1\right) \epsilon_{aij} x_j / r^2 \\ \phi &= (r \coth r - 1) x_a / r^2 \end{aligned}\right\} \qquad (8.5)$$

Monopoles of charge k with group $SU(2)$ can all be constructed from a particular algebraic curve of genus $(k-1)^2$, called the <u>spectral curve</u>. There are two approaches to this. The first is via twistor theory which starts from the observation that a solution of the Bogomolny equations gives rise to a self-dual connection on $\mathbb{R}^4$:

$$A_1 dx_1 + A_2 dx_2 + A_3 dx_3 + \phi dx_4 .$$

The second is the approach of Nahm, which uses the ADHM construction of §3 with matrices replaced by differential operators on an interval. It leads to the Nahm equations:

$$\left.\begin{aligned} \frac{dA_1}{dt} &= [A_2, A_3] \\ \frac{dA_2}{dt} &= [A_3, A_1] \\ \frac{dA_3}{dt} &= [A_1, A_2] \end{aligned}\right\}$$

for $k \times k$ matrix functions $A_i(t)$. This is a completely integrable system which can be linearized on the Jacobian of the spectral curve.

As in the case of instantons, explicit forms of solving the equations do not necessarily give the required information on the qualitative behaviour of their solutions. Here, however, it is quite easy to show that the only monopoles of charge 1 are in fact translates in $\mathbb{R}^3$ of the basic monopole (8.5) centred at the origin. The moduli space of 1-monopoles is thus just $\mathbb{R}^3$ itself.

There is a fundamental difference here between instantons and monopoles. The five dimensions of the moduli space of 1-instantons consisted of a choice of centre and a choice of <u>scale</u>. With monopoles we only have the choice of centre - there is no scale parameter.

This observation is a reflection of a more general analytical fact, proven in [JT], which is that monopoles of any charge are of <u>finite size</u> in the sense that there are uniform bounds (independent of A and ϕ) for $|F_A|$, $|\phi|$ and their derivatives. Thus unlike instantons, monopoles cannot become concentrated. This has an important bearing on the geometrical structure of the moduli space as we shall see next.

§9. Moduli spaces of monopoles

The necessary analysis for proving the existence and smoothness of a moduli space for the equivalence classes of monopoles under the action of the group of gauge transformations is modelled on that for instantons but requires considerably more effort owing to the non-compactness of $\mathbb{R}^3$. The relevant estimates and Fredholm operators which replace the usual elliptic estimates on compact manifolds were proved by Taubes [T]. Note that non-trivial monopoles cannot exist on compact manifolds, since the Bianchi identity gives

$$d_A{*}d_A\phi = d_A F_A = 0$$

and then integration by parts yields

$$\int_M |d_A\phi|^2 = \int (d_A{*}d_A\phi, \phi) = 0 .$$

Thus, whatever replacement for $\mathbb{R}^3$ we choose to define monopoles upon must necessarily be non-compact and hence will lead to more complicated analysis. Using the results of Taubes one can show that the moduli space of $SU(2)$ monopoles of charge k on $\mathbb{R}^3$ is a smooth manifold of dimension $4k - 1$. We have seen what this is for $k = 1$, namely $\mathbb{R}^3$ itself. More generally, any monopole has a centre and the $(4k-1)$-dimensional moduli space is a product:

$$\mathbb{R}^3 \times M^{4k-4}$$

where the first factor determines the centre.

The Higgs field ϕ of a monopole of charge k has in general k zeros. For each unoriented axis there is a unique axi-symmetric monopole, in which case the zeros coincide. In the case $k = 2$, one finds that, up to translation and rotation, a monopole is determined by the distance between the zeros of the Higgs field. This yields $4.2 - 1 = 7$ parameters - 3 for translation, 3 for rotation and one for the distance apart.

A more detailed analysis (see [AH]) shows that each 2-monopole has three axes of symmetry under rotation by π:

(1) the main axis, which becomes the axis of rotational symmetry for the axi-symmetric monopoles.

(2) the Higgs axis, which is the line joining the zeros of the Higgs field, and

(3) the third axis, orthogonal to (1) and (2).

This remark shows that the symmetry group of a monopole of charge 2 is a conjugate of the group Γ of diagonal matrices in $SO(3)$. The orbit structure of the 4-dimensional moduli space of centred 2-monopoles is then

$$SO(3)/\Gamma \times \mathbb{R}^+ \quad \cup \quad \mathbb{RP}^2$$

↑

distance between Higgs zeros — axially symmetric monopoles

A smooth model for this moduli space together with the $SO(3)$ action is then obtained by taking the 5-dimensional space of 3×3 symmetric real matrices A of trace zero and considering the unit sphere $\mathrm{Tr}A^2 = 1$. This contains two degenerate orbits of $SO(3)$ corresponding to matrices with two equal positive eigenvalues and two equal negative eigenvalues. Removing one of these gives a manifold diffeomorphic in an $SO(3)$-equivariant manner to the moduli space. (From a differential geometric point of view, the real projective planes are Veronese surfaces).

The monopole moduli spaces have well-defined L^2-metrics which are complete. This is basically a consequence of Uhlenbeck's theorem and the fact monopoles have a finite size and do not become concentrated. The L^2-metric on the space of centred monopoles (or alternatively on a circle bundle over the $(4k-1)$-dimensional moduli space) also has the property that it is hyperkähler.

This property is again due to the moduli space being, formally speaking, a hyperkähler quotient. The situation is analogous to that considered in §7. We consider the space of all L^2 pairs $(A,\phi) \in \mathcal{A} \times \Omega^0(\mathbb{R}^3;g)$ as an infinite-dimensional hyperkähler manifold with an action of a group $\mathcal{G}$ of gauge transformations with suitable behaviour at infinity. The hyperkähler moment map is then just

$$\underline{\mu}(A,\phi) = (F_{23}-d_1\phi, F_{31}-d_2\phi, F_{12}-d_3\phi)$$

and so the moduli space of solutions to the Bogomolny equations is a hyperkähler quotient.

The metric has one obvious further property - an $SO(3)$ action by isometries arising from the physical action on $\mathbb{R}^3$. There are two possible types of hyperkähler manifolds on which $SO(3)$ acts - those for which the representation on the 3-dimensional space of covariant constant 2-forms is trivial and those for which it is non-trival. In this case we have a non-trivial action - SO(3) rotates ω_1, ω_2 and ω_3, the three Kähler forms.

What we have then, for $k = 2$, is a 4-dimensional hyperkähler manifold with an SO(3) action which rotates the Kähler forms. This information enables us to circumvent the difficult task of directly computing the L^2-metric in the style of Groissier and go instead to a differential equation which describes all such metrics [GP] : the metric can be put in the form

$$g = (abc)^2 d\eta^2 + a^2\sigma_1^2 + b^2\sigma_2^2 + c^2\sigma_3^2 \tag{9.1}$$

where σ_1, σ_2, σ_3 are basic left-invariant 1-forms on SU(2) and a, b, c are functions of η, satisfying

$$\frac{2}{a}\frac{da}{d\eta} = (b-c)^2 - a^2 \quad \text{etc.}$$

The extra condition of <u>completness</u> gives the following unique exact solution:

$$\left.\begin{aligned} ab &= -2k(k'^2)K\frac{dK}{dk} \\ bc &= ab - 2(k'K)^2 \\ ca &= ab + 2(kK)^2 \end{aligned}\right\} \tag{9.2}$$

$$\eta = \frac{-K'}{\pi K} + \tfrac{1}{2}$$

where K, k, k' are the standard variables associated to elliptic functions

$$K(k) = \int_0^{\pi/2} \frac{d\phi}{\sqrt{1-k^2\sin^2\phi}}$$

$$K'(k) = K(k') \quad \text{and} \quad k^2 + k'^2 = 1 .$$

Of course, the appearance of elliptic integrals is not surprising, since we know that each 2-monopole gives a spectral curve of genus $(2-1)^2 = 1$ i.e. an elliptic curve. An interesting point to note is that the natural parameter η is the ratio of two periods of the elliptic curve, i.e. the natural coordinate in the upper halfplane. A derivation of this formula may be found in [AH].

One of the reasons for calculating this metric is to attempt to give examples of the dynamics of two monopoles. The basic idea we appeal to, enunciated by Manton, is the assumption that "slowly moving solutions to time-dependent equations are approximated by geodesic motion on the space of static solutions". In this context it means that we should be approximating solutions to the full time-dependent Yang-Mills-Higgs equations based on the Yang-Mills-Higgs functional in Minkowski space by geodesics on the metric we have just given. The situation where it really works is the motion of a ball bearing around the base of a bowl.

A full study of the geodesic motion for the metric (9.1) and (9.2) requires the use of powerful computing techniques and some recent work by Temple-Raston shows regions of chaotic behaviour as well as others which are more regular. Some examples of geodesics can however be seen immediately.

The first is to consider the fixed point set of Γ, the diagonal group of matrices of the form

$$\begin{pmatrix} \pm 1 & & \\ & \pm 1 & \\ & & \pm 1 \end{pmatrix}$$

with determinant 1. In the model with symmetric matrices this is the curve of diagonal matrices. Being the fixed point set of a group of isometries, it is totally geodesic, and since it is a curve it is actually a geodesic. The physical picture arising from this geodesic is of two monopoles colliding head on and then scattering at right angles. As we shall see later, this is a general phenomenon and does not require an explicit computation to reveal it.

The second type of geodesic, discovered by Bates and Montgomery, is a closed one, but a very special one, namely an orbit of a circle subgroup of $SO(3)$. This can be found by using a familiar principle in variational theory: if we are looking for a G-invariant critical point of a G-invariant functional, we need only consider the functional restricted to G-invariant points. In our situation we are therefore looking for S^1-invariant circles (i.e. orbits) and trying to minimize the energy of these to get a geodesic.

Now each orbit lies in an $SO(3)$ orbit and the metric (9.1) on one of these orbits is for constants a, b, c,

$$a^2\sigma_1^2 + b^2\sigma_2^2 + c^2\sigma_3^2 \, .$$

The energy here is critical for circles which point in the directions of principal axes of this quadratic form. However, it can be shown that a', b' and c' are non-zero, hence if the principal axes are distinct we can increase or decrease the energy of this circle by varying η. We must therefore have $a^2 = b^2$ (say). The angle the circle makes with the directions σ_1 and σ_2 can then be calculated by variation

$$2aa'\cos^2\theta + 2bb'\sin^2\theta = 0 .$$

The outcome is a geodesic consisting of two monopoles rotating about an axis which makes an angle of approximately $\pi/9$ with the Higgs axis

More details of other geodesics and the physical interpretation in terms of dyons can be found in the book [AH].

§10. Gauge theory and Riemann surfaces

So far, we have considered gauge theories and associated Yang-Mills-type equations in dimensions 4 and 3. It is natural to see also what the gauge theory point of view produces in 2 dimensions - on a Riemann surface. A notable application of these ideas is the paper of Atiyah & Bott [AB], who considered the straight Yang-Mills functiona

$$\int_M |F_A|^2 *1$$

on the space of connections over a compact Riemann surface M. Here the critical points lead to the equation

$$d_A * F_A = 0$$

i.e. the section $*F_A \in \Omega^0(M;g)$ is covariant constant. This is a very strong condition, and locally is uninteresting. Globally, however, one can define interesting moduli spaces which are ultimately moduli spaces of flat G connections. Using the theorem of Narasimhan and Seshadri which identifies the moduli space of stable holomorphic bundles on M with the moduli space of flat unitary connections, Atiyah and Bott obtain topological and geometrical information about this moduli space from the Yang-Mills viewpoint.

There is a generalization of this set-up which is more in keeping with the self-duality equations in $\mathbb{R}^4$ and the Bogomolny equations in $\mathbb{R}^3$. As we remarked in §8, a solution (A,ϕ) of the Bogomolny equations gives a self-dual connection on $\mathbb{R}^4$ by setting

$$A = A_1dx_1 + A_2dx_2 + A_3dx_3 + \phi dx_4 .$$

Alternatively, we can say that solutions of the Bogomolny equations are the self-dual connections on $\mathbb{R}^4$ which are invariant under x_4-translation. Similarly we can consider self-dual connections on

$\mathbb{R}^4$ which are invariant under translation in the x_3- and x_4-directions. These give us connection forms A_1, A_2 on $\mathbb{R}^2$ with coordinates (x_1,x_2) and Higgs fields (in the adjoint representation) ϕ_1 and ϕ_2 such that

$$A = A_1dx_1 + A_2dx_2 + \phi_1dx_3 + \phi_2dx_4$$

is self-dual connection. Spelling this out gives the equations

$$\left.\begin{aligned} F_A &= [\phi_1,\phi_2] \\ \nabla_1\phi_1 + \nabla_2\phi_2 &= 0 \\ \nabla_1\phi_2 - \nabla_2\phi_1 &= 0 \end{aligned}\right\} .$$

If we now set $dz = dx_1 + idx_2$ and $\Phi = \frac{1}{2}(\phi_1 - i\phi_2)dz$, then the equations take the form

$$\left.\begin{aligned} F_A + [\Phi,\Phi^*] &= 0 \\ d_A''\Phi &= 0 \end{aligned}\right\} . \qquad (1.1)$$

We call these the <u>self-duality equations</u> in two dimensions. Put in this form the equations are <u>conformally invariant</u> and can therefore be considered on a compact Riemann surface. Certainly solutions exist, for we have only to take $\Phi = 0$, $F_A = 0$ and look at flat unitary connections, so we can also consider the moduli space of equivalence classes of all solutions under the action of the group of gauge transformations. This, in the case of $SU(2)$, is the basis of the paper [H] to which we refer for more details.

The first result is that the moduli space is, except at the reducible solutions, a <u>smooth manifold</u> of dimension $4(g-1)\dim G$, where $g > 1$ is the genus of the Riemann surface on which the solutions are defined.

The second result (to be expected when we note that the dimension is divisible by four) is that the natural L^2-metric is <u>hyperkähler</u>. The point to note about this is that the metric here depends only on the conformal structure of Riemann surface M, for it arises ultimately via a hyperkähler quotient from the L^2-metric on $\mathcal{A} \times \Omega$ where $\mathcal{A}$ is the space of G-connections and $\Omega = \Omega^0(M;\mathfrak{g} \otimes_{\mathbb{C}} K)$ is the space of C^∞ sections of the adjoint bundle twisted with the canonical bundle (of forms of type $(1,0)$ on M) and here the metrics on $\mathcal{A}$ and Ω are defined by the $*$-operator on 1-forms, which is conformally invariant. To be more precise, we have

$$g(\dot{A},\dot{B}) = \int_M \text{trace}(\dot{A}\wedge *\dot{B}) \quad \text{on} \quad A, \quad \text{and}$$

$$g(\dot{\Phi},\dot{\Phi}) = \int_M \text{trace}(\dot{\Phi}\ \dot{\Psi}^*) \quad \text{on} \quad \Omega.$$

There are other basic properties of this metric as follows:

(1) If $G = SO(3)$ and $w_2 \neq 0$, then the metric is complete. This is again a manifestation of the general fact that solutions to these self-duality equations on Riemann surfaces do not become concentrated.

(2) There is an isometric action of the circle on the moduli space, given by

$$(A,\Phi) \to (A,e^{i\theta}\Phi) \ .$$

Note that this action preserves the equation (10.1).

(3) This circle action preserves one of the complex structures (call it I) of the hyperkähler metric. This fact can be used to give a Morse theoretic method of calculating the Betti numbers of the moduli space [H].

The self-duality equations (10.1) may be viewed as equations relating a pair of objects. On the one hand the d_A'' operator defines us a holomorphic structure on some associated vector bundle V, and then the second part of the equations gives us a holomorphic section Φ of $\text{End}\, V \otimes K$. The first part of the equation is a constraint for a certain metric (or equivalently connection) defined on V. This constraint imposes a constraint on the pair (V,Φ) which is called stability. We make the definition:

Definition: (V,Φ) is stable if for any Φ-invariant subbundle W, $\deg W/\text{rk}\, W < \deg V/\text{rk}\, V$. (10.2)

Here Φ-invariant means that $\Phi(W) \subset W \otimes K$. In the case $\Phi = 0$, this is the usual definition of stability for vector bundles over Riemann surfaces.

This notion of stability fits in with the equations because of the following theorem:

Theorem: (V,Φ) is stable iff there is a hermitian metric on V such that $F_A + [\Phi,\Phi^*] = 0$. (10.3)

This theorem was proved in the rank 2 case by the author [H] and in much greater generality (using a version of the Uhlenbeck-Yau theorem) by Simpson. Noting that $\Phi \to e^{i\theta}\Phi$ is holomorphic in this interpretation, we have a corollary:

<u>Corollary</u>: The moduli space, with complex structure I, is naturally the moduli space of stable pairs (V,Φ) .

This latter moduli space has been defined purely using algebraic geometry by Nitsure.

At this point one might ask: "How do we know that this is the right definition of stability?" We could equally ask the question for the Uhlenbeck-Yau theorem. This is generally the easy direction in such theorems: given the special connection, prove stability. In general, too, these properties are proven by <u>vanishing theorems</u>. The argument, for the stability condition (10.2), goes as follows in the case where $\operatorname{rk} V = 2$, $\deg V = 0$. Suppose that we have a solution of the equations $d''_A\Phi = 0$ and $F_A + [\Phi,\Phi^*] = 0$ and suppose, for a contradiction, that there is a line bundle $L \subset V$ of degree ≥ 0 which is Φ-invariant. First we choose a metric on the Riemann surface M itself with area 2π, then one can easily show that there exists a metric connection on L with curvature $(\deg L)\omega$ where ω is the volume form of M. Use this to put a metric on $L^* \otimes V$, and hence obtain a connection B.

The inclusion of L in V defines a holomorphic section s of $L^* \otimes V$ and we now consider

$$d''_B d'_B s + d'_B d''_B s = F(B)s .$$

Now $d''_B s = 0$ and $F(B) = F_A - (\deg L)\omega$, so

$$d''_B d'_B s = -[\Phi,\Phi^*]s - (\deg L)\omega s. \tag{10.4}$$

Now s is Φ-invariant, so locally $\Phi s = \lambda s$. This means that

$$\begin{aligned}\langle(\Phi\Phi^*-\Phi^*\Phi)s,s\rangle &= \langle\Phi\Phi^*s,s\rangle - |\lambda|^2\langle s,s\rangle \\ &= \langle(\Phi^*-\bar{\lambda})s \, , (\Phi-\lambda)s\rangle \geq 0 .\end{aligned}$$

Consequently, integrating (10.4) by parts,

$$\begin{aligned}0 \leq \int_M \langle d'_B s, d'_B s\rangle &= \int_M \langle d''_B d'_B s, s\rangle \\ &= \int \langle -[\Phi,\Phi^*]s - \deg L\omega s, s\rangle \leq 0\end{aligned}$$

and so all the integrands vanish, which forces $\deg L = 0$ and the solution to be reducible: $d_B s = 0$, $\Phi\Phi^* = \Phi^*\Phi$.

Let us consider now some examples of stable pairs, and the solutions to the self-duality equations they give rise to.

(1) Let V be a stable bundle and $\Phi = 0$. Then the theorem says that $F_A = 0$, i.e. V admits a flat unitary connection compatible with the holomorphic structure - this is the theorem of Narasimhan & Seshadri. Note that it is also the first case of the Uhlenbeck-Yau theorem.

(2) Let $K^{\frac{1}{2}}$ denote a <u>square-root</u> of the canonical bundle (i.e. a spin-structure on M), and set $V = K^{\frac{1}{2}} \oplus K^{-\frac{1}{2}}$. This vector bundle on its own is definitely not stable, but if we set

$$\Phi = \begin{pmatrix} 0 & 0 \\ 1 & 0 \end{pmatrix}$$

then the only <u>Φ-invariant</u> subbundle of V is $K^{-\frac{1}{2}}$ and this is stable if its degree $-(g-1)$ is less than zero. We assume then that $g > 1$. (Note that we interpret "1" as the canonical element in $\mathrm{Hom}(K^{\frac{1}{2}};K^{-\frac{1}{2}}) \otimes K$.)

Out of this holomorphic data, Theorem (10.3) throws back a solution to the self-duality equations (10.1). There is actually a uniqueness part to the equations which gives more information. We note that if

$$g = \begin{pmatrix} i & 0 \\ 0 & -i \end{pmatrix} \quad \text{then} \quad g^{-1}\Phi g = -\Phi$$

and g is an automorphism of V. Now if (A,Φ) is a solution of the self-duality equations so is $(A,-\Phi)$. The uniqueness tells us that g is an automorphism of the <u>connection</u> A, and hence A is reducible to a hermitian connection on $K^{\frac{1}{2}}$, (which is just a metric h on the surface M itself) with curvature F.

The equation $F_A + [\Phi,\Phi^*] = 0$ now becomes

$$F = h$$

or in other words a metric of <u>constant curvature -4</u>. We therefore have an alternative proof of the <u>uniformization theorem</u> for compact Riemann surfaces.

(3) The equivalence class of the solution above is actually invariant under the full circle action on the moduli space: a fixed point is in general either a flat connection or a decomposable one: $V = L \oplus L^*$, with Φ of the form

$$\Phi = \begin{pmatrix} 0 & 0 \\ a & 0 \end{pmatrix}, \quad a \in H^0(M;L^{-2}K) .$$

Here $\deg L > 0$ for stability so the holomorphic section a has $2g - 2 - 2 \deg L$ zeros. Conversely given such a divisor we can define (up to equivalence) a stable pair (V,Φ). Thus the fixed-point set of the circle action in the moduli space is a symmetric product of M with itself. In fact, considering $SO(3)$ bundles instead of $SU(2)$ ones, we can find the k-fold symmetric product S^kM for $k < 2g - 2$ lying as a totally geodesic submanifold of the moduli space. It inherits a canonical Kähler metric this way, though we have insufficient knowledge to calculate this.

§11. Flat connections and Teichmüller space

We just gave one description of the moduli space of solutions to the self-duality equations on a Riemann surface. It was the equivalence classes of stable pairs (V,Φ) and had complex structure I. We had already noticed, however, that this moduli space had a hyperkähler structure and hence other complex structures J & K. We shall now describe the moduli space referring to the complex structure J.

To begin this, suppose (A,Φ) satisfies the self-duality equations (10.1), then it is easy to see that the non-unitary connection

$$d_A + \Phi + \Phi^* \tag{11.1}$$

is flat. (In fact the self-duality equations are equivalent to the statement that $d_A + \zeta\Phi + \zeta^{-1}\Phi^*$ is flat for all $\zeta \in \mathbb{C}^*$) .

If we start out with a solution for the compact simple group G with complex group G^c, then this is a flat G^c connection. In fact a vanishing theorem shows that if the solution is irreducible then the flat connection (11.1) is irreducible in the sense that the holonomy lies in no proper parabolic subgroup. This is a "stability" statement, and we have a converse theorem proved by Donaldson [D4] in the rank 2 case and Corlette in general:

Theorem: Every irreducible flat G^c connection is equivalent to a flat connection of the form $d_A + \Phi + \Phi^*$ where (A,Φ) satisfies the self-duality equations. (11.2)

This is the analogue of (10.3) for the complex structure J. A corollary of this is that the moduli space is isomorphic to the space of equivalence classes of flat G^c connections. This space can be defined independently of the conformal structure of the Riemann surface M. It is simply

$$\mathrm{Hom}(\pi_1(M);G^c)^{irr}/G^c$$

and is a Zariski open set in an affine variety.

As an example of the flat connection obtained this way we can take the solution to the self-duality equation corresponding to a metric of constant negative curvature as in §10. Here the flat connection has holonomy contained in $SL(2,\mathbb{C})$. In fact the uniformization of the surface gives us an action of π_1 on the upper half-plane and hence a homomorphism into $PSL(2,\mathbb{R})$. A lifting of this to $SL(2,\mathbb{R})$ is the flat connection (11.1).

Now <u>any</u> conformal structure on the Riemann surface M can be uniformized and hence gives rise to a representation of π_1 into $SL(2,\mathbb{R})$. Consequently from Theorem (11.2) above there must correspond a solution to the self-duality equations. We shall see next how the moduli space of constant negative curvature metrics under diffeomorphisms homotopic to the identity - <u>Teichmüller space</u> - appears from this observation.

We start with the familiar holomorphic bundle $V = K^{\frac{1}{2}} \oplus K^{-\frac{1}{2}}$ on M and now take

$$\Phi = \begin{pmatrix} 0 & q \\ 1 & 0 \end{pmatrix} .$$

Here $q \in H^0(M;\mathrm{Hom}(K^{-\frac{1}{2}};K^{\frac{1}{2}}) \otimes K) = H^0(M;K^2)$ is a <u>quadratic differential</u>. The special case $q = 0$ gave us the constant curvature metric associated to the underlying conformal structure on M. The same sort of argument shows that the corresponding point in the moduli space is again invariant under $\Phi \to -\Phi$ and consequently the connection A which one obtains by solving the self-duality equations is reducible. This gives a metric h on M compatible with the conformal structure but whose curvature is

$$F = (1-|q|^2)h . \qquad (11.3)$$

Now, however, we put

$$\hat{h} = q + h + \frac{q\bar{q}}{h} + \bar{q} .$$

This is a smooth section of the bundle $S^2T^* \otimes \mathbb{C}$ (i.e. a quadratic form on T) but which is also real. One can then check:

(1) $\hat{h}$ is a metric (i.e. it is positive definite). This comes down to the condition $|q|^2 < 1$ which is proved by appeal to the strong maximum principle.

(2) $\hat{h}$ has constant curvature -4. This is just a reinterpretation of equation (11.3).

(3) Any metric of constant curvature -4 is isometric to one of these metrics by a diffeomorphism homotopic to the identity. Here we can use a theorem in harmonic maps due to Eells, Earle and Sampson [E] which asserts that there is a unique harmonic diffeomorphism homotopic to the identity between any two surfaces of constant negative curvature. It is a well-known consequence of harmonic maps that the (2,0) part of the pulled-back metric is a holomorphic quadratic differential. Alternatively, we can appeal to Donaldson or Corlette's theorem, which are both theorems about harmonic maps.

These three facts give us from an unusual viewpoint the well-known result that Teichmüller space is contractible: here it is simply given by the space of quadratic differentials $q \in H^0(M;K^2)$, a vector space of complex dimension $(3g-3)$. This space is canonically the cotangent space to Teichmüller space at the given conformal structure on M, so the above construction gives an identification of Teichmüller space with the cotangent space at a point. This makes it analogous to the exponential map in differential geometry.

(We should note that in this model, the natural induced metric is not the Weil-Peterson metric although its Kähler form is remarkably enough the Weil-Peterson form.)

One extra corollary is the fact that Teichmüller space appears here as a component of the fixed point set of the involution $\Phi \to -\Phi$. When we refer to the picture of flat $SL(2,\mathbb{C})$ connections it consists of a component of the space of equivalence classes of flat $PSL(2,\mathbb{R})$ connections, in fact the ones for which the associated $\mathbb{RP}^1$ bundle has Euler class $(2g-2)$. This is a known result due to Goldman and others.

It has a generalization to an arbitrary simple Lie group, in fact. For the general linear group we take as a vector bundle

$$V = K^n \oplus K^{n-1} \oplus \ldots \oplus 1 \oplus K^{-1} \oplus \ldots \oplus K^{-n}$$

for a bundle of odd rank, or

$$V = K^{n/2} \oplus K^{(n-2)/2} \oplus \ldots \oplus K^{\frac{1}{2}} \oplus K^{-\frac{1}{2}} \oplus \ldots K^{-n/2}$$

for a bundle of even rank. For the Higgs field Φ we take a matrix such that

$$\left.\begin{aligned} \Phi_{ij} &= 0 && i > j+1 \\ &= 1 && i = j+1 \\ &= 0 && i = j \\ &= q_k && i = j+k \end{aligned}\right\}$$

where q_k is a holomorphic section of K^{k+1}. The dimension of the space of such Higgs fields is, from the Riemann-Roch theorem

$$\begin{aligned} \sum_{k=1}^{m} \dim_{\mathbb{C}} H^0(M;K^{k+1}) &= (g-1)\sum_{1}^{m-1}(2k+1) \\ &= (m^2-1)(g-1) \\ &= \dim PSL(m,\mathbb{R})\cdot(g-1) \ . \end{aligned}$$

Using similar methods to those above one can show that this is a component of the space of equivalence classes of flat irreducible $PSL(m,\mathbb{R})$ connections and is therefore a direct generalization of Techmüller space. Whether there is always a geometrical structure on M that it parametrizes is anybody's guess.

§12. Vortices and Skyrmions

We have looked in the previous sections at gauge theories in 2, 3 and 4 dimensions which have well-defined moduli spaces and whose topological and geometrical structure is by now quite well understood. The theories which gave rise to these moduli spaces had physical origins, but had to undergo some simplification in order to be mathematically tractable. The relevance of instantons or classical solutions to the Yang-Mills-Higgs equations in the Prasad-Sommerfield limit in understanding any reasonable physical phenomenon is questionable. Instead, we should regard them as models, exhibiting characteristic behaviour which we might expect to hold in more complicated, but more physical, situations.

There are two gauge theories which are tantalizingly close in many aspects to some of those we have considered, have far more physical relevance, but have unfortunately been far more resistant to attack by analytical methods. The first of these is a two-dimensional theory - the Abelian Yang-Mills-Higgs model including the quartic potential term (8.1) - which actually made its appearance as long ago as 1954 as the Ginzburg-Landau theory of superconductivity. The static solutions of these equations which are absolute minima of the functional (analogues of the monopoles of §8 and §9) are called vortices.

The second theory, which has been proposed as a mathematical model for the proton or neutron is the theory associated to the Skyrme functional, which gives rise to a 3-dimensional theory analogous to the theory of harmonic maps. We shall briefly study these two theories next and see the moduli space problems which they suggest.

We first consider vortices, associated to the 2-dimensional functional

$$ymH = \int_{\mathbb{R}^2} |F_A|^2 + |d_A\phi|^2 + (1-|\phi|^2)^2$$

where here the connection A is defined on a principal $U(1)$ bundle, equivalently a complex line bundle L, and ϕ - the Higgs field - is a section of L^2. Details of the analytical aspects of this situation can be found in Jaffe & Taubes' book [JT].

The boundary conditions for finite action imply $|\phi| \to 1$ as $r \to \infty$. Coupled with the decay of the curvature, this effectively defines a map

$$\mathbb{R}^2/B_R \to S^1$$

which has a degree k - the vortex charge or vortex number. Just as in the instanton or monopole situation this charge provides a lower bound for the functional. The argument is in fact close to the vanishing theorem manipulations related to the stability criterion in §10. We write

$$|d_A\phi|^2 = |d_A'\phi|^2 + |d_A''\phi|^2$$

$$= -*(d_A''d_A'\phi,\phi) + |d_A''\phi|^2 + 2d(d_A'\phi,\phi)$$

and then use the formula

$$d_A'd_A'' + d_A''d_A' = F_A$$

to give

$$|d_A\phi|^2 = *(d_A'd_A''\phi,\phi) + *2(F_A\phi,\phi) + |d_A'\phi|^2 + *d(d_A'\phi,\phi)$$

$$= 2|d_A''\phi|^2 + 2*F_A|\phi|^2 + *d\theta \qquad (12.1)$$

where $\int_{\mathbb{R}^2} *d\phi$ can be evaluated in terms of the asymptotic data by Stokes' theorem.

Using this rearrangement of terms we write

$$|F_A|^2 + (1-|\phi|^2)^2 = |F_A - *(1-|\phi|^2)|^2 + 2(F_A,*(1-|\phi|^2))$$

$$= |F_A - *(1-|\phi|^2)|^2 + 2*F_A - 2*F_A|\phi|^2 \,.$$

Evaluating the integrals of both $*d\theta$ and $*F_A = *dA$ in terms of the behaviour at infinity, one finds

$$\int_{\mathbb{R}^2} |F_A|^2 + (1-|\phi|^2)^2 + |d_A\phi|^2$$

$$= \int |F_A - *(1-|\phi|^2)|^2 + 2|d_A''\phi|^2 + 2\pi k$$

and so the action is bounded below by $2\pi k$ with equality if and only if the Abelian vortex equation is satisfied:

$$\left.\begin{array}{l} d_A''\phi = 0 \\ F_A = *(1-|\phi|^2) \end{array}\right\} \,. \qquad (12.2)$$

This is beguilingly similar to the special solution of the self-duality equations (11.3), together with the holomorphicity of the quadratic differential q, but there is an essential difference. Equation (12.2) involves $*1$, the volume 2-form of the flat metric on $\mathbb{R}^2$ and the curvature of the line bundle L. Equation (11.3) involves only one connection and metric.

The vortex equations are not particular solutions to the self-dual Yang-Mills equations in $\mathbb{R}^4$, and so we cannot expect the methods we have at our disposal there to apply. It is true that they may be interpreted as SO(3)-invariant solutions to the self-dual Yang-Mills equations on $S^2 \times \mathbb{R}^2$, but there again we have no special solution techniques to apply. Even the case of the rotationally symmetric 1-vortex solution is not known in finite terms.

What is known about vortices is their moduli space, which was essentially calculated by Jaffe & Taubes. What they proved was that given k unordered points in $\mathbb{R}^2$ (possibly with multiplicities) there exists an Abelian vortex solution whose Higgs field ϕ vanishes at just those points. The solution is, moreover, unique up to gauge equivalence. This means that the moduli space consists of the space of unordered k-tuples $S^k\mathbb{C}$, the k-th symmetric produce of $\mathbb{C}$. If we think of those points as the zeros of a monic polynomial

$$p(z) = z^k + a_, z^{k-1} + \ldots + a_1$$

then the moduli space is just the vector space $\mathbb{C}^k$ of coefficients of all such polynomials and so topologically very simple.

Vortices share many features with monopoles. One is that they are of finite size and cannot become concentrated. This is on the one hand a consequence of Jaffe and Taubes' analysis, and on the other visible in the fact that the moduli space of a single vortex is $\mathbb{R}^2$ - its only parameter is its location. The second is that each monopole has a centre - the centre of mass of the Higgs zeros. Thirdly, for each centre there is a unique axially symmetric vortex configuration - the polynomial $p(z) = z^k$.

When it comes to the definition of the L^2-metric on the vortex moduli space, then strictly speaking we need to construct a suitable elliptic theory to make the moduli space a manifold - we don't know a priori that the parameters $(a_1, \ldots, a_k)$ are smooth coordinates. However, assuming this can be done, we expect a metric to be defined and we can then also consider the scattering of vortices, as with monopoles. Here the 90° scattering phenomenon (which we saw for monopoles) in a direct collision becomes clear. Let us consider the moduli space of two centred vortices. This is the 1-complex dimensional space of quadratic polynomials of the form

$$p(z) = z^2 - a \quad (a \in \mathbb{C})$$

and the "locations" of the vortices are the two points $z = \pm\sqrt{a} \in \mathbb{C}$.

Consider a geodesic tangential at $a = 0$ to the real axis. We can approximate this curve linearly by

$$a = t \qquad (t \in \mathbb{R}) .$$

When $t > 0$, the two vortices $\pm\sqrt{t}$ lie on the real axis, but after collision t is less than zero and then $\pm\sqrt{t}$ lie on the orthogonal imaginary axis - they part after collision at right angles to their initial impact.

One feature to note here is also that $\frac{dz}{dt} = \frac{1}{2}t^{-\frac{1}{2}}$ becomes infinite at $t = 0$, so one might have doubts about the vortices "moving slowly" - the rationale behind the approximation of solutions to time-dependent equations by geodesic motion. However computer calculations for both monopoles and vortices show that the action density for an axially symmetric solution has a toroidal or circular distribution which keeps away from the axis of symmetry. When we are in this "non-particle" region we have an uncertainty in determining the "location" of the particles - is it where the action density is a maximum or where the Higgs field vanishes? Which is it that is "slowly moving"? This perhaps provides an answer to the paradox.

Finally, let us turn to the Skyrme model. In its simplest version we consider a smooth map

$$f : \mathbb{R}^3 \to S^3$$

and its differential df, considered as a map of tangent spaces:

$$df : T_x\mathbb{R}^3 \to T_{f(x)}S^3 .$$

There is an induced map on the exterior powers of the corresponding spaces, in particular

$$\Lambda^2 df : \Lambda^2 T_x\mathbb{R}^3 \to \Lambda^2 T_{f(x)}S^3 .$$

The <u>Skyrme energy</u> of the map is defined as

$$E = \int_{\mathbb{R}^3} |df|^2 + |\Lambda^2 df|^2 \tag{12.3}$$

where we use the flat metric on $\mathbb{R}^3$ and the constant curvature metric on S^3 to evaluate the norms. The first term in this expression is the usual harmonic map functional. The extra term is natural in a 3-dimensional context bearing in mind the Hodge duality between 1-forms and 2-forms: the Skyrme energy is "balanced" with respect to the *-operator in the same way that the harmonic map functional is balanced for 2-manifolds and the Yang-Mills functional for 4-manifolds.

The analysis of this functional is facilitated by considering the <u>strain tensor</u>:

$$D = (df)^*df .$$

This is a positive semi-definite self-adjoint transformation on the tangent bundle of $\mathbb{R}^3$, with eigenvalues $\lambda_1^2, \lambda_2^2, \lambda_3^2$. In terms of these eigenvalues, the energy is

$$E = \int_{\mathbb{R}^3} \lambda_1^2 + \lambda_2^2 + \lambda_3^2 + \lambda_1^2\lambda_2^2 + \lambda_2^2\lambda_3^2 + \lambda_3^2\lambda_1^2$$

$$= \int_{\mathbb{R}^3} (\lambda_1 - \lambda_2\lambda_3)^2 + (\lambda_2 - \lambda_3\lambda_1)^2 + (\lambda_3 - \lambda_1\lambda_2)^2 + 6\int_{\mathbb{R}^3} \lambda_1\lambda_2\lambda_3 \ .$$

This latter term is a topological invariant - essentially the degree of the map f when extended to the compactification S^3 of $\mathbb{R}^3$. We have then a lower bound

$$E \geq 12\pi^2(\deg f) \ .$$

Our experience with instantons, monopoles and vortices would suggest we consider the case

$$\lambda_1 = \lambda_2\lambda_3; \quad \lambda_2 = \lambda_3\lambda_1; \quad \lambda_3 = \lambda_1\lambda_2$$

where equality occurs, but this quickly leads to a condition which is impossible - f must be an isometry! In fact for degree 1, the minimum energy configuration has energy approximately $1.23 \times 12\pi^2$, somewhat higher than the topological lower bound of $12\pi^2$. (See [M] for details of Skyrmions).

Here we have no special self-duality equations or Bogomolny equations to solve, and the minima appear from numerical work to be isolated and symmetrical. There is thus no obvious moduli space to generalize the ones we are by now familiar with. Nevertheless, Manton has suggested the existence of a 12-parameter space of Skyrmion configurations endowed with an L^2-metric, and this time a potential function, too. The geodesic motion coupled to the potential should describe a scattering process. We are left here with the open problem of determining in some manner what this finite-dimensional submanifold of the infinite-dimensional space of configurations should be.

13. Moduli spaces in their own right

The question raised by the vortex and Skyrmion equations is: "Do we have to know precisely the configurations in order to discover the geometry of the moduli space?" Put another way, we might consider a general class of geometries and then as a secondary question consider which ones can be identified as moduli spaces. As an example, it is perfectly natural to consider the geometry of the upper half plane and the action of $SL(2;\mathbb{Z})$ on it. It may ultimately make it easier for us to relate each point with an elliptic curve, but there is no particular reason to start from that point of view. As another example, it is natural to consider the class of principally polarized Abelian varieties in its own right, and then to ask the subsidiary question of which ones are moduli spaces of line bundles on Riemann surfaces - The Schottky problem, which has only recently been solved.

Adopting this point of view, one fact we can observe is that many of the moduli spaces we have encountered are in fact spaces of holomorphic maps. The vortex case, where the moduli space is a space of polynomials, is the simplest example, but equally the moduli space of $SU(2)$ monopoles of charge k has such an interpretation. This is the theorem of Donaldson:

Theorem: The space of based rational maps of degree k from $\mathbb{C}P^1$ to $\mathbb{C}P^1$ is diffeomorphic to a circle bundle over the moduli space of $SU(2)$ monopoles of charge k.

The maps in question may be written in the form

$$f(z) = \frac{a_0+a_1z+\ldots+a_{k-1}z^{k-1}}{b_0+b_1z+\ldots+b_{k-1}z^{k-1}+z^k}$$

$$= p(z)/q(z) \qquad (13.1)$$

where the polynomials have no common factor. This is an open set in the complex 2k-dimensional affine space of pols $(p(z),q(z))$. The property of no common factor introduces the complicated topology into the space. The circle action is simply multiplication by $e^{i\theta}$.

In the case of a single monopole the map is

$$f(z) = \frac{a_0}{b_0+z}$$

where $b_0 \in \mathbb{C}$ and $a_0 \in \mathbb{C}^*$.

The moduli space is thus $\mathbb{C} \times \mathbb{C}^* \cong \mathbb{R}^3 \times S^1$ and given this observation we can begin to think in terms of a "position" of a particle in $\mathbb{R}^3$ and a "phase" in S^1 as parameters, without reference to any particular equations.

If we considered simply a configuration of k such particles, the moduli space would be the symmetric product $S^k(\mathbb{C} \times \mathbb{C}^*)$ which is, however, a singular space. Remarkably, the space of rational maps provides a natural resolution of the singularities of this space [AH]: this space is in the first place smooth, but we also have a surjective map to $S^k(\mathbb{C} \times \mathbb{C}^*)$ which is generically one-to-one.

To define this, we take a rational function $f = p(z)/q(z)$ as in (13.1) and associate to it the unordered set of k points $(z_i, p(z_i))$ $(1 \le i \le k)$ where $z_i \in \mathbb{C}$ are the roots of $q(z)$. Since p and q have no common factors, $p(z_i)$ is indeed an element of $\mathbb{C}^*$.

It is possible, then, to view the space of rational functions from a particular angle and see the relationship with the moduli space of particle - like objects in 3-space.

The moduli space of instantons on $\mathbb{R}^4$ of charge k has a similar interpretation, due to the theorem of Atiyah & Donaldson [A2]:

Theorem: The moduli space of instantons on $\mathbb{R}^4$ of charge k with group G is diffeomorphic to the space of based holomorphic maps from $\mathbb{C}P^1$ to ΩG, the space of loops on G.

A possible candidate for the Skyrmion space might be found by an approach of this sort : finding a suitable space of holomorphic maps which perhaps relates to a desingularization of configurations of points and phases. There is no a priori reason why such moduli spaces should be identified with spaces of maps, but we have parallel theories to guide us.

If we begin to think of moduli spaces as being geometrical structures in their own right, as having metrics and function theories of an especially interesting nature and only incidentally consider them as parametrizing solutions to Yang-Mills type equations, then we may be able to see more clearly their mathematical role. It is just possible that they are the 20th-century versions of Jacobians, whose final structure has only now begun to be understood.

References

[Am] S. Amari, "Differential geometric methods in statistics", Lecture Notes in Statistics 28, Springer Verlag, Heidelberg (1985).

[A1] M.F. Atiyah, "Geometry of Yang-Mills fields", Lez. Fermi Acc. Naz. dei Lincei of Scuola Norm. Sup., Pisa (1979).

[A2] M.F. Atiyah, "Instantons in two and four dimensions", Commun. Math. Phys. 93 (1984), 437-451.

[AB] M.F. Atiyah & R. Bott, The Yang-Mills equations over Riemann surfaces, Phil. Trans. R. Soc. Lond. A308 (1982), 523-615.

[AH] M.F. Atiyah & N.J. Hitchin, "The geometry and dynamics of magnetic monopoles", Princeton Univ. Press, Princeton (1988).

[AHS] M.F. Atiyah, N.J. Hitchin & I.M. Singer, "Self-duality in four-dimensional Riemannian geometry", Proc. Roy. Soc. Lond. A362 (1978) 425-461.

[ADHM] M.F. Atiyah, V.G. Drinfeld, N.J. Hitchin & Yu. I. Manin, "Construction of instantons", Phys. Lett. 65A (1978) 18 185-187.

[B] N.P. Buchdahl, "Instantons on $\mathbb{C}P^2$", Commun. Math. Phys. 24 (1986) 19-52.

[D1] S.K. Donaldson, "Anti-self-dual Yang-Mills connections over complex algebraic surfaces and stable vector bundles", Proc. London Math. Soc. 50 (1985), 1-26.

[D2] S.K. Donaldson, "Vector bundles on the flag manifold and the Ward correspondence", in "Geometry of Today", eds. E. Arbarello, C. Procesi, E. Strickland, Birkhauser, Boston (1985) 109-119.

[D3] S.K. Donaldson, "Infinite determinants, stable bundles and curvature", Duke Math. J. 54 (1987) 231-247.

[D4] S.K. Donaldson, "Harmonic maps and the self-duality equations", Proc. London Math. Soc 55 (1987) 127-132.

[D5] S.K. Donaldson, "Polynomial invariants for smooth four-manifolds", preprint Oxford (1988).

[DMM] H. Doi, Y. Matsumoto & T. Matumoto, "An explicit formula of the metric on the moduli space of BPST-instantons over S^4", "A Fête of Topology", Academic Press (1987).

[E] C.J. Earle & J. Eells, "A fibre bundle description of Teichmüller theory", J. Diff. Geom. 3 (1969) 19-43.

[FU] D.S. Freed & K.K. Uhlenbeck, "Instantons and four-manifolds", Springer Verlag, New York (1984).

[GP] G. Gibbons & C. Pope, "The positive action conjecture and asymptotically Euclidean metrics in quantum gravity", Commun. Math. Phys. 66 (1979) 287-290.

[Gr] D. Groissier, "The geometry of the moduli space of $\mathbb{C}P^2$ instantons", preprint Stony Brook (1988).

[GrP] D. Groissier & T.H. Parker, "The Riemannian geometry of the Yang-Mills moduli space", Commun. Math. Phys. 112 (1987) 663-689.

[Ha] L. Habermann, "On the geometry of the space of Sp(1) instantons with Pontrjagin index 1 on the 4-sphere", Annals. of Global Analysis & Diff. Geom. (to appear).

[H] N.J. Hitchin, "The self-duality equations on a Riemann surface", Proc. London Math. Soc. 55 (1987), 59-126.

[HKLR] N.J. Hitchin, A. Karlhede, U. Lindström & M.Rocek,"Hyperkähler metrics and supersymmetry", Commun. Math. Phys. 108 (1987) 535-589.

[JT] A. Jaffe & C.H. Taubes, "Vortices and monopoles", Birkhauser, Boston (1980).

[K] K. Kodaira, "Complex manifolds and deformation of complex structures", Springer Verlag, New York (1986).

[Ku] K. Kuranishi, "On the locally complete famlies of complex analytic structures", Annals of Math. 75 (1962),536-477.

[M] N.S. Manton, "Geometry of Skyrmions", Commun. Math. Phys. 111 (1987) 469-478.

[MQ] V. Mathai & D. Quillen, "Superconnections, Thom classes and equivariant differential forms", Topology 25 (1986) 85-110.

[Q] D. Quillen, "Determinants of Cauchy-Riemann operators over a Riemann surface". Funct. Anal. Appl. 19 (1985) 31-34.

[T] C.H. Taubes, "Stability in Yang-Mills theories", Commun. Math. Phys. 91 (1983) 235-263.

[UY] K. Uhlenbeck & S.T. Yau, "Hermitian-Einstein connections in stable vector bundles", Commun. Pure Appl. Math. Suppl. 39(S) (1986) 257-293.

[W] E. Witten, "Topological quantum field thoery", Commun. Math. Phys. 117 (1988) 353-386.

TOPICS IN CONFORMAL FIELD THEORY AND STRING THEORY

L. Alvarez-Gaumé
Theory Division
CERN, CH-1211 Geneva 23, Switzerland

LECTURE 1

The aim of these lectures is to review some recent developments in string theory [1]. In particular we will present a formalism which can deal in principle with conformally invariant field theories on arbitrary Riemann surfaces. As explicit examples, we will discuss the bosonic string in flat space and the superstring. Since we will not assume the audience to be acquainted with conformal field theories in $d = 2$ dimensions, this lecture and part of the next will review some of the relevant features of these theories.

Conformally invariant field theories describe the long-range behaviour of correlation functions in systems undergoing a second-order phase transition. From the pioneering work of Wilson, Kadanoff, Fisher, Polyakov, etc. [2], we have learned to appreciate the fundamental rôle of scale invariance in the renormalization group approach to critical phenomena. Briefly, we start with some microscopic Hamiltonian $H_o[\phi_i^{(0)}, a]$ defined on some lattice of lattice spacing a, with some fields $\phi_i^{(0)}$ and coupling constants $g_i^{(0)}$. To describe long-range phenomena, we want to thin out the fast degrees of freedom (integrate them out). For example, we can define "block-spin" variables on a lattice of spacing 2a. Integrating out the fast modes gives a new Hamiltonian $H_1[\phi_i^{(1)}, g_i^{(1)}, 2a]$ with new fields and coupling constants. If we keep on performing the renormalization group transformation $RH_i = H_{i+1}$, we get a picture very similar to a dynamical system whose configuration space is the space of Hamiltonians, and the flow is defined by the renormalization group transformation. Starting with some initial Hamiltonian, the renormalization group flow produces a trajectory, and at each point we obtain an effective Hamiltonian describing the physics at a particular length scale. One of the characteristic features of a critical system is that the correlation length is effectively infinite compared with the initial lattice spacing, implying that these systems are described by the fixed points of the renormalization group transformations $RH^* = H^*$. The critical system is scale invariant. Thus, close to criticality, the most convenient way of describing the physics is in terms of scaling fields. They are eigenvectors of the scale transformation.

In the language of CFT, these fields are the primary fields, and their dimensions (scaling dimensions) determine the critical exponents of a critical system, the heat capacity as a function of T: $C_V \sim (T-T_c)^{-\alpha}$, the spontaneous magnetization, etc.

Since close to the critical point the system has "forgotten" the original lattice structure, we can describe its correlation functions in terms of a local field theory. The order parameters are represented by some local fields ϕ_i, all with well-defined scaling dimensions and, at criticality, the interactions are supposed to be scale invariant:

$$\xi_i \to \lambda \xi_i \qquad \phi(\xi_i) \to \lambda^{-\Delta_i} \phi_i(\lambda^{-1}\xi) \tag{1.1}$$

Scale invariance automatically implies conformal invariance in a local 2-d QFT. Our exposition of conformal field theory will follow closely the work of Belavin, Polyakov and Zamolodchikov [3].

Given an action S in the presence of external gravity, the energy-momentum tensor is defined by the response of the system to a gravitational field:

$$\Theta_{\mu\nu} = \frac{\delta S}{\delta g^{\mu\nu}} \frac{1}{\sqrt{g}} \Big|_{g=\delta} \tag{1.2}$$

$$P^\mu = \int d^{n-1}x \, \theta^{0\mu} \qquad J^{\mu\nu} = \int d^{n-1}x \, (x^\mu \theta^{0\nu} - x^\nu \theta^{0\mu}) \tag{1.3}$$

θ generates arbitrary co-ordinate transformations:

$$\int j_\mu(V) \, d\Sigma^\mu = \int V_\mu \theta^{\mu\nu} d\Sigma_\nu \tag{1.4}$$

$j_\mu(V)$ is conserved if $\partial_\mu \theta^{\mu\nu} = 0$ and

$$\partial_\mu (V_\nu \theta^{\nu\mu}) = \frac{1}{2} (\partial_\mu V_\nu + \partial_\nu V_\mu) \theta^{\nu\mu} = 0 \tag{1.5}$$

hence

$$\partial_\mu V_\nu + \partial_\nu V_\mu = 0 \qquad (\text{Killing conditions}) \tag{1.6}$$

For dilatations, $V^\mu = x^\mu$, and invariance implies

$$\theta_\mu{}^\mu = 0 \tag{1.7}$$

An infinitesimal conformal transformation satisfies

$$\partial_\mu V_\nu + \partial_\nu V_\mu - \frac{2}{d} g_{\mu\nu} \partial_\alpha V^\alpha = 0 \tag{1.8}$$

or its covariant version:

$$ds^2 = g_{\mu\nu} dx^\mu dx^\nu$$

$$g_{\mu\nu}(x') dx'^\mu dx'^\nu = \Omega^2(x) g_{\mu\nu} dx^\mu dx^\nu \tag{1.9}$$

If V_μ satisfies $\partial_\mu V_\nu + \partial_\nu V_\mu - (2/d) g_{\mu\nu} \partial_\alpha V^\alpha = 0$, and $\theta_\mu{}^\mu = 0$, then we have full conformal invariance:

$$\partial_\mu (V_\nu \theta^{\nu\mu}) = \partial_\mu V_\nu \, \theta^{\nu\mu} =$$

$$= \frac{1}{2} \left(\partial_\mu V_\nu + \partial_\nu V_\mu - \frac{2}{d} g_{\mu\nu} \partial_\alpha V^\alpha \right) \theta^{\nu\mu} = 0 \tag{1.10}$$

hence $\theta_\mu{}^\mu = \partial_\mu \theta^{\mu\nu} = 0$ implies conformal symmetry. If $d > 2$, the conformal group for Minkowski space $M_{p,q}$ with signature $(+,\overset{p}{\ldots},+,\ -,-,\overset{q}{\ldots},-)$ is isomorphic to some covering of SO(p+1, q+1). Two dimensions, however, give a richer structure. In light-cone co-ordinates

$$\partial_+ V_+ = \partial_- V_- = 0 \tag{1.11}$$

and the other condition is empty:

$$\partial_+ V_- + \partial_- V_+ - \frac{2}{2} (\partial_+ V_- + \partial_- V_+) \equiv 0$$

with

$$ds^2 = dx^+ dx^- \qquad g_{+-} = g_{-+} = 1/2$$

$$g^{+-} = g^{-+} = 2$$

$$x^{\pm} = x^0 \pm x^1 \qquad \partial_{\pm} = \frac{1}{2}\left(\frac{\partial}{\partial x^0} \pm \frac{\partial}{\partial x^1}\right) \tag{1.12}$$

Then a general conformal transformation in two dimensions takes the form

$$x'^{+} = f(x^{+}) \quad , \quad x'^{-} = g(x^{-}) \tag{1.13}$$

and the light cone does not change. To avoid infra-red problems, we will compactify space and work on $S^1 \times R$. Under Wick rotation,

$$\begin{aligned} x^0 + x^1 &\rightarrow -i\,w \\ x^0 - x^1 &\rightarrow -i\,\bar{w} \end{aligned} \tag{1.14}$$

Furthermore, if we conformally map the cylinder onto a plane minus the origin:

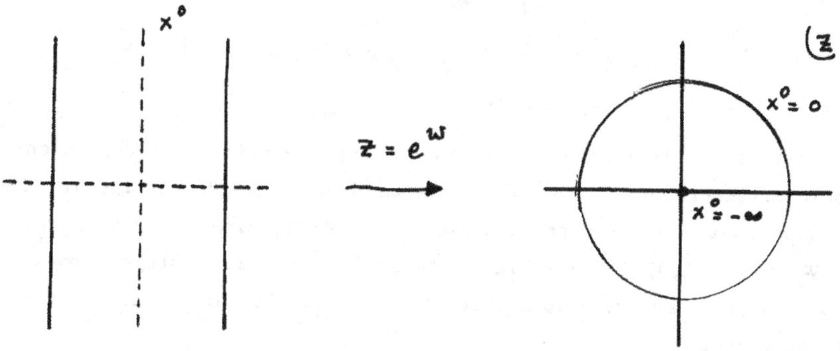

Fig. 1

Quantizing the theory on the Z-plane is known as radial quantization. Circles centred at the origin correspond to equal-time surfaces. This is particularly useful because it makes scale invariance very explicit. The Möbius group

$$z \rightarrow \frac{az+b}{cz+d} \qquad a,b,c,d \in \mathbb{C}, \quad ad-bc = 1 \tag{1.15}$$

acts on $C \cup \{\infty\}$, and scale transformations

$$z \rightarrow \lambda z \quad , \quad \bar{z} \rightarrow \bar{\lambda}\bar{z} \tag{1.16}$$

are equivalent to time translation in the original co-ordinates.

Since we have an infinite number of symmetries, we have an infinite number of conservation laws. These are neatly summarized by saying that θ_{zz} ($\theta_{\bar{z}\bar{z}}$) is a holomorphic (antiholomorpic) field

$$\bar{\partial}_z \theta_{zz} = 0$$
$$\partial_z \theta_{\bar{z}\bar{z}} = 0$$

We can expand $T(z) \equiv \theta_{zz}(z)$ (and similarly for $\bar{T}$) in a Laurent series:

$$T(z) = \sum_{n \in \mathbb{Z}} L_n z^{-n-2} \tag{1.17}$$

T has scaling dimension 2, and the conserved quantities are

$$L_n = \oint_0 \frac{dz}{2\pi i} z^{n+1} T(z)$$

$$v(z) = \sum_n \epsilon_n z^{n+1} \tag{1.18}$$

Conformal invariance is manifest when the objects we write down depend only on the holomorphic structure of the surface where we perform the computations. The vector $v(z)$ behaves under conformal transformations as $\partial/\partial z$, since (naively) $T(z)$ behaves as dz^2, $V\cdot T$ is a (1,0) holomorphic differential, and its integral over a closed contour only depends on the holomorphic structure of the surface and on the homology class of the contour.

The ∞-dimensional symmetry algebra $\{L_n\}$ is a representation of the algebra of vector fields

$$v_n = -z^{n+1}\frac{d}{dz} \qquad [v_n, v_m] = (n-m)\, v_{n+m} \tag{1.19}$$

$T(v) \equiv \oint v(z)T(z)$ gives a representation of this algebra on the Hilbert space of states of the field theory. This representation is projective (see below). Now t-ordering becomes radial ordering. To compute the (anti) commutators, we only need to know the operator product expansions (OPEs) of the operators involved. In $d = 2$, these infinite dimensional algebras are very useful in statistical mechanics. They often determine completely the spectrum of anomalous dimensions. Since the scaling fields diagonalize L_0, some among them can be chosen to behave like tensors of type $(h,\bar{h})$:

$$\phi_{h,\bar h}(z,\bar z)\, dz^h d\bar z^{\bar h} \qquad (h-\bar h \in \mathbb{Z} \text{ or } \mathbb{Z}+1/2) \tag{1.20}$$

Then

$$\phi_{h,\bar h}(f(z))\, f'(z)^h\, dz^h = \phi_h(z)\, dz^h$$

(same for antiholomorphic parts which we omit). Then, for infinitesimal transformations:

$$z \to z + v(z)$$
$$\delta_\epsilon \phi(z) = [\,T(\epsilon), \phi(z)\,] = (v\partial + h\partial v)\,\phi(z) \tag{1.21}$$

hence

$$\delta_n \phi(z) = [\,L_n, \phi(z)\,] = \left(z^{n+1}\frac{d}{dz} + h(n+1)z^n\right)\phi(z) \tag{1.22}$$
$$v_n = z^{n+1}$$

Fields transforming as in (1.22) are known as primary fields. There is a distinguished subalgebra in (1.21) represented by the global holomorphic vector fields on the sphere $\{1, z, z^2\}$ (conformal Killing vectors). They are the generators of the Möbius group. Hence L_{+1}, L_0, L_{-1} represent the generators of this group on the Hilbert space of the theory. Since this group $[SL_2(C)]$ is the conformal symmetry of the compactified plane, we can postulate the existence of a unique state $|0\rangle$ in the Hilbert space $\mathcal{H}$ which is $SL_2(C)$-invariant:

$$L_{\pm 1}|0\rangle = L_0|0\rangle = 0 \tag{1.23}$$

In an operator language this is equivalent to saying that the operator algebra contains the unit operator. The equal-time commutators of the energy-momentum tensor can be computed using the OPE. This is based on the "contour swapping" argument familiar from the early days of string theory. Given two holomorphic operators $A(z)$, $B(w)$, we can compute the commutator:

$$[\,A(f), B(g)\,]_{E.T.}$$

$$A(f) = \oint_{C_0} f(z)A(z) \qquad B(g) = \oint_{C_0} g(z)B(z) \tag{1.24}$$

(from now on the contour integral contains the factor $1/2\pi i$) with the contours taken as circles centred at the origin. In radial quantization, (1.24) can be represented as

$$[A(f),B(g)] = \oint_{C_1} f(z)A(z) \oint_{C_2} g(w)B(w) - \oint_{C_2} g(w)B(w) \oint_{C_1} f(z)A(z)$$
$$|z|>|w| \qquad\qquad |w|>|z|$$

and the contours are represented in Fig. 2.

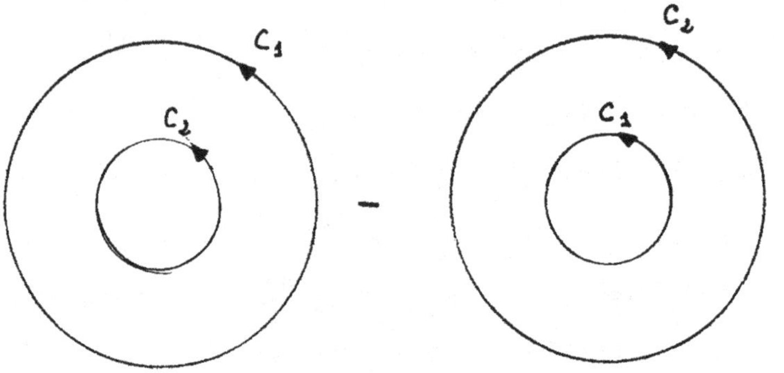

Fig. 2

Leaving z fixed in the first integral in (1.24), we can deform the C_2 contour past C_1. This will produce two terms: one cancels the second double integral, and the other is a small contour of z around w followed by an integral of w around the origin, namely

$$[A(f),B(g)] = \oint_0 dw\, g(w) \oint_w A(z)B(w)f(z) \tag{1.25}$$

By analyticity of A and B, we can choose the z-contour in (1.25) as small as we please, and therefore the z-integral will be completely determined by the singular terms of the OPE between A and B [as long as f(z) is holomorphic in a neighbourhood of w]. This information, together with (1.21), implies that for primary fields:

$$T(z)\phi(w) = \frac{h}{(z-w)^2}\phi(w) + \frac{1}{z-w}\partial\phi(w) + \text{analytic} \tag{1.26}$$

For the energy-momentum tensor itself, we obtain (1.26) with h = 2 plus an extra term, because for unitary theories the two-point function of T(z) cannot vanish. In the plane it is completely determined by dimensional arguments and the fact that we compute the correlation function on the SL_2-invariant ground state

$$\langle T(z)T(w)\rangle = \frac{c/2}{(z-w)^4} \tag{1.27}$$

where c is a constant depending on the theory considered. This, together with (1.26), yields

$$T(z)\,T(w) = \frac{c/2}{(z-w)^4} + \frac{2}{(z-w)^2}\,T(w) + \frac{1}{z-w}\,\partial T(w) + \ldots \qquad (1.28)$$

Using (1.18) and (1.25), we finally obtain the Virasoro algebra

$$[L_n, L_m] = (n-m)\,L_{n+m} + \frac{c}{12}\,(n^3-n)\,\delta_{n+m,0} \qquad (1.29)$$

This is a central extension of the algebra of vector fields (1.19), and it justifies the claim made previously that T(v) provides a projective representation of (1.19) on the Hilbert space $\mathcal{H}$. Notice first of all that for n = ±1,0, the central term in (1.29) vanishes. The representation theory of (1.29) is very rich, and some of its highlights are the minimal models of Belavin, Polyakov and Zamolodchikov [4] and the unitary series of Friedan, Qiu and Shenker [3].

The highest-weight representations of (1.29) are characterized by a highest-weight state

$$L_n\,|h\rangle = 0 \quad n>0 \qquad L_0\,|h\rangle = h\,|h\rangle \qquad (1.30)$$

and all the states in the representation or Verma module V(c,h) can be written in the form

$$L_{-n_1} \cdots L_{-n_N}\,|h\rangle \qquad (1.31)$$

with L_0 eigenvalue $h + \sum_{i=1}^{N} n_i$. The Verma module is completely characterized by (c,h). We can equivalently characterize the states in V(c,h) in terms of operators. The highest-weight conditions (1.30) are equivalent to having a primary field $\phi_h(z)$ in the operator algebra. The other states (1.31), known as descendants, can be represented as

$$(L_{-n_1} \cdots L_{-n_N}\,\phi_h)(z) = \oint_{C_1} \frac{T(z_1)}{(z_1-z)^{n_1-1}} \oint_{C_2} \frac{T(z_2)}{(z_2-z)^{n_2-1}} \cdots$$

$$\cdots \oint_{C_N} \frac{T(z_N)}{(z_N-z)^{n_N-1}}\,\phi(z) \qquad (1.32)$$

$$C_1 \supset C_2 \supset \ldots \supset C_N \supset z$$

and they have rather complicated conformal transformation. Using the analyticity properties of $T(z)$, it is easy to derive the conformal Ward identities corresponding to the insertion of one or more energy-momentum tensors in a correlation function involving primary fields. For example the correlator

$$\langle 0| T(z)\,\phi_1(z_1) \dots \phi_n(z_n)|0\rangle$$

for fixed $z_1, \dots, z_n$ is a holomorphic function of z in $S^2-\{z_1, \dots, z_n\}$. The pole structure of this function as $z \to z_i$ is determined by the OPE of $T(z)$ with $\phi_i(z_i)$. We thus conclude

$$\begin{aligned}\langle 0| T(z)\,\phi_1(z_1) \dots \phi_n(z_n)|0\rangle &= \\ &= \sum_{i=1}^{n} \left(\frac{h_i}{(z-z_i)^2} + \frac{1}{z-z_i}\frac{\partial}{\partial z_i} \right) \langle 0|\phi_1(z_1) \dots \phi_n(z_n)|0\rangle \end{aligned} \tag{1.33}$$

and similarly for more insertions of $T(z)$. A simple conclusion one obtains from (1.32) and (1.33) is that the correlation functions of descendant fields are completely determined by those of the primary fields, and therefore the latter are the fundamental quantities to evaluate in a Conformal Field Theory.

The central term in (1.29) also allows us to compute the transformation law of $T(z)$ under arbitrary conformal transformations

$$T(z) \to \left(\frac{d\zeta}{dz}\right)^2 T(\zeta) + \frac{c}{12}\{\zeta, z\} \tag{1.34}$$

where $\{\zeta, z\}$ is the Schwarzian derivative:

$$\{\zeta, z\} = \frac{\zeta'''}{\zeta'} - \frac{3}{2}\left(\frac{\zeta''}{\zeta'}\right)^2 \tag{1.35}$$

hence in the transformation from the cylinder to the plane $\zeta = e^z$

$$\{e^z, z\} = -1/2 \tag{1.36}$$

and only L_0 changes:

$$(L_0)_{cyl.} = (L_0)_{plane} - \frac{c}{24} \tag{1.37}$$

This change is important in the modular properties of Virasoro characters. The derivation of (1.34) can be done in a number of ways. Perhaps the simplest appears in Ref. [5], where the properties of the energy-momentum tensor in a conformal theory are derived in the presence of an arbitrary geometry on the plane.

LECTURE 2

We have so far studied the "kinematics" of conformal invariance in two dimensions. We concluded that the operator algebra of the theory contains the identity and the energy-momentum tensor, and all other operators fall into conformal classes $[\phi_{h,\bar{h}}]$, each furnishing a representation of the Virasoro algebra fully characterized by c and $(h,\bar{h})$. The existence of an SL_2-invariant ground state completely determines the form of the two- and three-point functions in a conformal theory. More generally, for an n-point function of the primary fields $\phi_1, \ldots, \phi_n$, it is easy to show that the Green function has the general form

$$\langle \phi_1(z_1)\ldots\phi_n(z_n)\rangle = \Big(\prod_{i<j} z_{ij}^{-\gamma_{ij}}\Big) f(\eta_a)$$

$$\sum_{j\neq i}\gamma_{ij} = 2h_i \qquad \gamma_{ij}=\gamma_{ji}\ , \qquad z_{ij}=z_i-z_j \tag{2.1}$$

and η_a are n-3 independent harmonic ratios formed with the n points $z_1, \ldots, z_n$. For a four-point function there is only one such variable:

$$\eta = \frac{z_{12}\, z_{34}}{z_{13}\, z_{24}}$$

The two-point function becomes

$$\langle \phi_{h,\bar{h}}(z,\bar{z})\,\phi_{h',\bar{h}'}(w,\bar{w})\rangle = \delta_{h,h'}\delta_{\bar{h},\bar{h}'}(z-w)^{-2h}(z'-w')^{-2\bar{h}} \tag{2.2}$$

and only a constant needs be determined on the three-point function:

$$\langle \phi_i(z_i)\,\phi_j(z_j)\,\phi_k(z_k)\rangle = C_{ijk}\, z_{ij}^{-\gamma_{ij}}\, z_{jk}^{-\gamma_{jk}}\, z_{ik}^{-\gamma_{ik}}$$
$$\gamma_{ij} = h_i+h_j-h_k \quad \text{etc.} \tag{2.3}$$

The coefficients C_{ijk} are the basic structure constants of the operator algebra, and together with the spectrum of dimensions $(h,\bar{h})$ and c they determine the theory.

Strong constraints on C_{ijk} are imposed by the requirement of associativity of the OPE [3]. The set of fields of the theory $\{A_i(x)\}$ fall into conformal families (Verma modules). The OPE between two primary fields ϕ_i, ϕ_j can be written as:

$$\phi_i(z,\bar{z})\,\phi_j(0,0) = \sum_p \sum_{\{k,\bar{k}\}} C_{ij}{}^p\, z^{h_p-h_i-h_j+\sum k} \cdot \bar{z}^{\bar{h}_p-\bar{h}_i-\bar{h}_j+\sum \bar{k}}\, \beta_{ij}^{p\{k\}}\, \bar{\beta}_{ij}^{p\{\bar{k}\}} \left(L_{-k_1}\cdots L_{-k_N}\, \bar{L}_{-\bar{k}_1}\cdots \bar{L}_{-\bar{k}_N}\,\phi_p\right)(0,0) \tag{2.4}$$

where p runs over primary fields and the coefficients $\beta,\bar{\beta}$ can be determined from c, h_i, h_j, h_k using conformal invariance [4]. Defining

$$|h\rangle = \lim_{z,\bar{z}\to 0} \phi_{h,\bar{h}}(z,\bar{z})\,|0\rangle$$

$$\langle h| = \lim_{z,\bar{z}\to\infty} \langle 0|\,\phi_{h,\bar{h}}(z,\bar{z})\, z^{2L_0}\,\bar{z}^{2\bar{L}_0} \tag{2.5}$$

The implications of associativity of the OPE are best expressed in terms of the four-point function

$$\langle k|\,\phi_\ell(1,1)\,\phi_n(x,\bar{x})\,|m\rangle \tag{2.6}$$

Substituting (2.4) in the operator product between ϕ_n and ϕ_m, and using (2.3), we obtain

$$\langle k|\,\phi_\ell(1,1)\,\phi_n(x,\bar{x})\,|m\rangle = \sum_p C_{nm}{}^p\, C_{k\ell p}\, \mathcal{F}_{nm}^{\ell k}(p|x) \cdot \overline{\mathcal{F}_{nm}^{\ell k}(p|x)} \tag{2.7}$$

where the multivalued analytic function $\mathcal{F}$ is known as a conformal block, normalized so that

$$\mathcal{F}_{nm}^{\ell k}(p|x) \underset{x\to 0}{=} x^{h_p-h_\ell-h_k}\,(1+\ldots) \tag{2.8}$$

This block can be represented graphically as in Fig. 3.

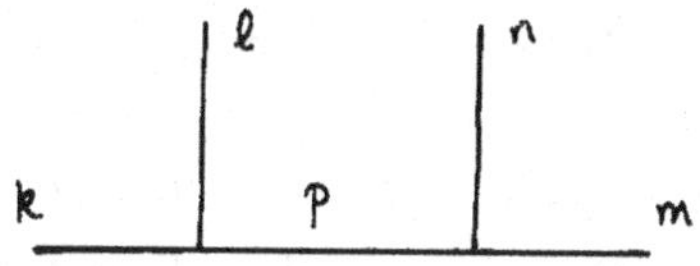

Fig. 3

If we want to use associativity of the OPE, instead of performing first the OPE between n and m and then with ℓ, we can use SL_2 invariance first and then do the OPE between ℓ and n. Associativity then implies [4]:

$$G^{\ell k}_{nm}(x,\bar{x}) = G^{mk}_{n\ell}(1-x, 1-\bar{x}) \tag{2.9}$$

Graphically this is similar to crossing in ordinary field theory (Fig. 4):

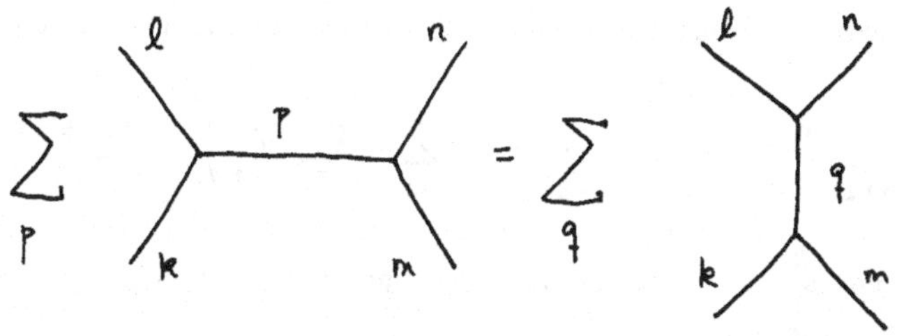

Fig. 4

When (2.9) is expressed in terms of the conformal blocks (2.8), we obtain a non-linear equation involving the structure constants $C_{ij}^{\ k}$, the dimensions $h,\bar{h}$ and the central term c. The classification of conformal field theories (or universality classes in statistical mechanics) is closely related to the classification of solutions to these equations. In some cases the structure constants $C_{ij}^{\ k}$ and correlation functions of primary fields can be computed due to the presence of null-vectors on the Verma modules. Given some module V(c,h), there is a null-vector at level N if there is a state $|\chi\rangle$ in V(c,h) satisfying

$$L_n|\chi\rangle = 0 \quad n>0\,, \qquad L_0|\chi\rangle = (h+N)|\chi\rangle \tag{2.10}$$

It is easy to show that $|\chi\rangle$ is orthogonal to all of V(c,h) and to the states in any other module; therefore it can be safely set to zero, $\chi = 0$. This yields non-trivial differential equations for the correlation functions involving the primary field ϕ_h. For example, if h satisfies

$$h = \frac{1}{16}\left(5-c \pm \sqrt{(c-1)(c-25)}\right) \tag{2.11}$$

then the state

$$|\chi\rangle = \left(L_{-2} - \frac{3}{2(2h+1)} L_{-1}^2\right)|h\rangle \tag{2.12}$$

is null. From the conformal Ward identities we can derive the equation:

$$\left(\frac{3}{2(2h+1)}\frac{\partial^2}{\partial z^2} - \sum_{i=1}^{n}\frac{h_i}{(z-z_i)^2} - \sum_{i=1}^{n}\frac{1}{z-z_i}\frac{\partial}{\partial z_i}\right)\langle\phi_h(z)\phi_1\ldots\phi_n\rangle = 0 \tag{2.13}$$

Kac [6] has given a list of all reducible representations V_h. They are labelled by pairs of positive integers (n,m). Writing

$$\begin{aligned} c &= 1 - 24\alpha_0^2 \\ \alpha_+ + \alpha_- &= 2\alpha_0 \\ \alpha_+\alpha_- &= -1 \end{aligned} \tag{2.14}$$

the module generated by a field of dimension

$$h(n,m) = -\alpha_0^2 + \frac{1}{4}\left(n\alpha_+ + m\alpha_-\right)^2 \tag{2.15}$$

has a null-vector at level nm yielding a differential equation of order nm (because the null-vector will contain a term proportional to $L_{-1}^{nm}\,\phi$, and using the Ward identities, we can replace L_{-1} by $\partial/\partial z$) for correlation functions involving the field $\phi_{n,m} = \phi_{h(n,m)}$. Among all the theories satisfying (2.15), those with the extra requirement

$$\alpha_+/\alpha_- = -q/p$$

q,p being relatively prime positive integers, have

$$c = 1 - 6(p-q)^2/pq$$

$$h(n,m) = \frac{1}{4pq}\left[(mp-nq)^2 - (p-q)^2 \right] \qquad (2.16)$$

and the operator algebra truncates to a finite number of conformal families $0 < n < p$, $0 < m < q$ [4]. If one requires the theory to be unitary, it was shown in Ref. [3] that $q = p+1$, $p > 2$. The differential equations induced by the null-vector in (2.16) can all be solved in terms of a Coulomb gas representation [7].

The OPE coefficients determine the fusion rules of the theory. They give the number of independent ways of coupling three conformal families:

$$[\phi_i] \times [\phi_j] = \sum_k N_{ij}{}^k\, [\phi_k] \qquad (2.17)$$

Equivalently, $N_{ij}{}^k$ counts the number of different three-point vertices in Fig. 5.

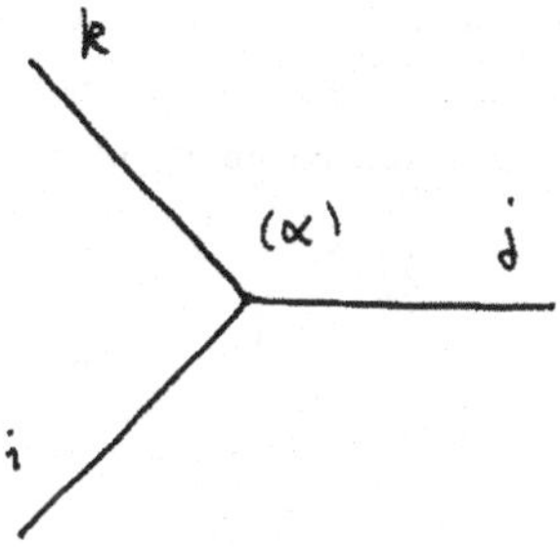

Fig. 5

The associativity of the OPE implies that the matrices $(N_i)_j{}^k \equiv N_{ij}{}^k$ commute:

$$N_i N_j = N_j N_i$$
$$N_i N_j = N_{ij}{}^k N_k \qquad (2.18)$$

hence we have a commutative associative algebra whose structure constants are positive integers. One of the more interesting recent developments in conformal field theory is the analysis by E. Verlinde of this algebra [8], which implies that the fusion rules (2.18) are diagonalized by the matrix S representing the modular transformation $\tau \to -1/\tau$ on the characters $\chi_i(\tau)$ of the modules appearing in the theory.

This claim was checked in many examples in Ref. [8], and it was proved by Moore and Seiberg [9] in a paper where they wrote down a set of polynomial equations for Rational Conformal Field Theories. A conformal theory is rational if it is characterized by a symmetry algebra containing 1 and the Virasoro algebra at least, and with a finite number of primary fields with respect to this algebra.

The fusion algebra (2.18) is useful in determining the number of conformal blocks contributing to a given correlation function. For instance, the number of blocks contributing to the four-point function in Fig. 6 is:

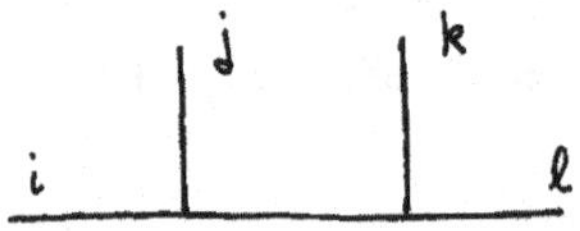

Fig. 6

$$N_{ijkl} = \sum_{p} N_{ij}{}^{p} N_{pk}{}^{l} \tag{2.19}$$

The associativity of (2.18) implies that $N_{ijk\ell}$ is independent of what basis we choose to draw the blocks. We could instead have chosen the dual diagram in Fig. 7, and $N_{ijk\ell}$ would still be the same. For modular invariant theories this count of conformal blocks extends also to diagrams containing loops.

Fig. 7

For higher point states (Fig. 8), we obtain:

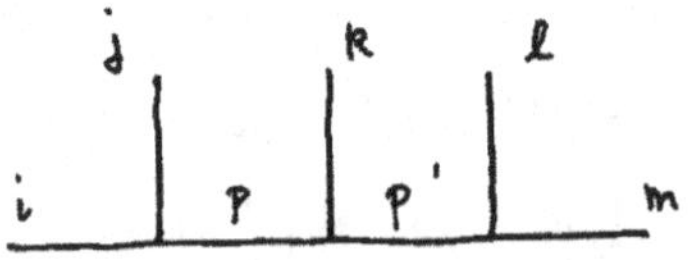

Fig. 8

$$N_{ijklm} = \sum_{p,p'} N_{ij}{}^{p} N_{pk}{}^{p'} N_{p'l}{}^{m} \tag{2.20}$$

etc. For a more general block $\mathcal{F}_I(z_1, \ldots)$, let I label the possible intermediate channels in the diagram. Then the physical correlation function is obtained by combining holomorphic and antiholomorphic blocks:

$$G(z_i,\bar{z}_j) = \sum_{I,J}{}' \mathcal{F}_I(z_i)\,\overline{\mathcal{F}_J(z_j)}\; g_{IJ} \tag{2.21}$$

into a local monodromy- and duality-invariant combination. From a geometrical point of view, the duality and monodromy transformations act on the blocks $\mathcal{F}_I$ and they form a representation of the braid group on $n = 1$ strands (for the n-point function). This feature brings in some interesting connection of conformal theories with the representation theory of braid groups, knots, polynomials, the Yang-Baxter equation, etc. (for a review and references, see [10]).

We will approach string theory from the point of view of conformal field theory in $d = 2$. The two dimensions represent the parameter space necessary to trace out the string history. In the remaining part of this lecture, we will give an overview of the Polyakov approach to the covariant quantization of the string (for many more details and references to the original literature, see the review in [11]). For a string moving in some target space-time, the amplitude corresponding to the process described in Fig. 9 is:

$$G(C_2,C_3;C_1) = \sum_{S,\,\partial S = C_1\cup C_2\cup C_3} e^{-\mathrm{Area}(S)} \tag{2.22}$$

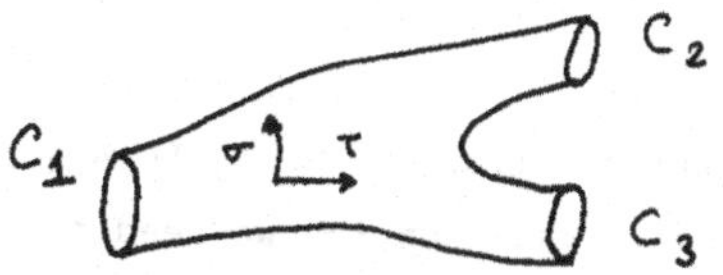

Fig. 9

The area in (2.22) is the area induced on the surface S by the target space metric.

Consider a bosonic string whose co-ordinates are represented by $X^{\mu}(\sigma,\tau)$. The induced metric is

$$G_{\mu\nu}(X)\,\partial_a X^{\mu}(\sigma,\tau)\,\partial_b X^{\nu}(\sigma,\tau) = \gamma_{ab}(\sigma,\tau) \tag{2.23}$$

For example, we can take $G_{\mu\nu} = \delta_{\mu\nu}$ (flat space). The quantization of this system is fairly complicated. It contains second-class constraints. Another procedure to quantize it, introduced by Polyakov, is to consider alternatively

$$S[X,g] = \frac{1}{2\pi\alpha'} \int d\sigma\, d\tau \sqrt{g}\, g^{ab}\, \partial_a X^\mu \partial_b X^\nu \eta_{\mu\nu} \tag{2.24}$$

Then:

$$\frac{\delta S}{\delta g^{ab}} = 0 \qquad \frac{\delta S}{\delta X_\mu} = \Box X^\mu = 0 \tag{2.25}$$

The first equation is a second-class constraint indicating the vanishing of the two-dimensional energy-momentum tensor:

$$\partial_a X^\mu \partial_b X^\nu \eta_{\mu\nu} - \frac{1}{2} g_{ab}\, g^{cd}\, \partial_c X^\mu \partial_d X^\nu \eta_{\mu\nu} = 0 \tag{2.26}$$

Solving for g_{ab} gives the induced metric. S is invariant under Diff(Σ) and Weyl rescalings. In the functional integral quantization, we have:

$$G(C';C) = \sum_{\text{topologies}} \sum_{\{g_{ab}\}} \frac{1}{\text{Vol}(\text{Diff}) \times \text{Vol}(\text{Weyl})} \sum_{(X^\mu)} e^{-S[X,g]} \tag{2.27}$$

Consistency requires that the field theory be conformal invariant with c = 26. This follows from the Faddeev-Popov quantization prescription. After a laborious procedure (see [11]), the measure contains a new set of anticommuting fields b_{zz}, c^z, $\bar{b}_{\bar{z}\bar{z}}$, $\bar{c}^{\bar{z}}$.

Choosing a slice of Metric/Diff×Weyl ≡ moduli space, where Metric ≡ space of metrics on a surface of genus g, the quantity we need to study is schematically

$$\sum_{\text{genera}} \lambda^g \int_{\mathcal{M}_g = \text{moduli space}} d\mu_g \int [dX]\, e^{-i \int \partial X \bar{\partial} X + \int (b \bar{\partial} c + \bar{b} \partial \bar{c})} \tag{2.28}$$

For two-dimensional surfaces, the space of metrics on a fixed topology modulo diffeomorphisms and conformal transformations is a finite-dimensional space. If we count degrees of freedom, a metric has three independent components. A diffeomorphism depends on two functions and a conformal transformation introduces one more arbitrary function. For the sphere, it is possible to show (see, for example, [12]) that we can always transform any metric to the standard one using this freedom. For a genus-g surface, the complex structure depends on a set of 6g-6 real parameters [12], and therefore the space $\mathcal{M}_g$ = Metrics/Diff×Weyl is a finite-dimensional space (the moduli space of curves of genus g).

The requirement of conformal invariance is easy to implement. A free scalar field X generates a conformal theory. The basic two-point functions are:

$$\langle \partial X^\mu(z)\, \partial X^\nu(w) \rangle = -\frac{\eta^{\mu\nu}}{(z-w)^2}$$

$$\langle X^\mu(z)\, X^\nu(w) \rangle = -\eta^{\mu\nu} \ln |z-w|^2 \tag{2.29}$$

The energy-momentum tensor is

$$T(z) = -\frac{1}{2}\, \partial X \cdot \partial X \tag{2.30}$$

Then

$$T(z)\,T(w) = \frac{d/2}{(z-w)^4} + \cdots \tag{2.31}$$

hence $c^{(x)} = d$. What is the central extension for the b-c system? In general, if b has spin j, and c spin 1-j:

$$b(z) = \sum_{n\in\mathbb{Z}} b_n z^{-n-j} \qquad c(z) = \sum_{n\in\mathbb{Z}} c_n z^{-n-1+j}$$

$$\{ b_n, c_m \} = \delta_{n+m,0}$$

$$b(z)\,c(w) = \frac{1}{z-w} + \cdots = c(z)\,b(w) \tag{2.32}$$

Under a co-ordinate transformation,

$$\delta_v b(z) = v\partial b(z) + j\partial v\; b(z)$$
$$\delta_v c(z) = v\partial c(z) + (1-j)\partial v\; c(z) \tag{2.33}$$

This implies that the energy-momentum tensor is

$$T(z) = j\partial c\, b - (1-j)\, c\partial b \tag{2.34}$$

Now the central extension is easy to compute:

$$c_j = -2\,(6j^2 - 6j + 1) \tag{2.35}$$

hence for $j = 2$, $c_2 = -26$. Thus

$$c^{(matter)} + c^{(gh)} = d - 26 = 0 \Rightarrow d = 26$$

However, we can achieve $c = 26$, by considering four uncompactified dimensions, and at the same time some conformal theory with $c = 22$. In the case of $d = 26$, the Polyakov amplitude at genus g is obtained by integrating a measure of the form

$$\frac{\rho \wedge \bar{\rho}}{(\det \operatorname{Im} \Omega)^{13}} \tag{2.36}$$

where ρ is the Mumford form in moduli space [13], and Ω is the period matrix of the surface. Next, we want to construct the Polyakov measure in a purely operatorial language. The advantage is that many of the manipulations can be done in the standard language of operators, and all the complications with gauge fixing are circumvented. Furthermore, in the formalism to be introduced in the next lectures, it is quite easy to obtain:

1) Differential equations for the measure.
2) Belavin-Knizhnik theorem.
3) Analysis of infinities.
4) Physical state conditions.
5) It is easy to show that the BRST operator Q_{BRST} acts as the exterior derivative d in moduli space $\mathcal{M}_g$, leading to the decoupling (up to boundary components in $\mathcal{M}_g$) of spurious BRST states.
6) It also clarifies the interpretation of the energy-momentum tensor as a connection on $\mathcal{M}_g$ and its action on moduli space.

LECTURE 3

In previous lectures, we worked directly in a two-dimensional space-time with the topology of a cylinder $S^1 \times R$. In this case (and in some cases also for the torus) we have the standard operator formalism which often simplifies the computations, and from a physicist's point of view it is fairly intuitive. It is a bit more difficult to construct an operator formalism on higher genus Riemann surfaces. Several groups have pursued the development of this formalism emphasizing different aspects. One started by solving some overlap equations [14] and eventually developed into the group theoretical approach to string amplitudes [15]; the Copenhagen group started with the N-reggeon vertex built in a BRST-invariant way mixed with the functional integral formulation [16,17]. In Ref. [18], a formalism inspired by string field theory is derived, and in Ref. [19] the starting point was inspired by possible models for the universal moduli space proposal of Friedan and Shenker [20], and exploits the more geometrical aspects of Riemann surface theory. I will report on work done in collaboration with Gomez, Reina, Moore, Nelson, Vafa and Sierra with regard to an operator formalism for strings and superstrings. In the original motivation, we wanted to have a formulation of CFT and in particular string theory on a higher genus surface, in such a way that all the information about the geometry and topology is coded on a particular state. More precisely, for the case of the plane, $z = 0$ ($\tau = -\infty$) and $z = +\infty$ ($\tau = +\infty$) are the "in" and "out" regions. Any correlation function can be written as

$$\langle 0 | \phi_1(z_1) \ldots \phi_N(z_N) | 0 \rangle \tag{3.1}$$

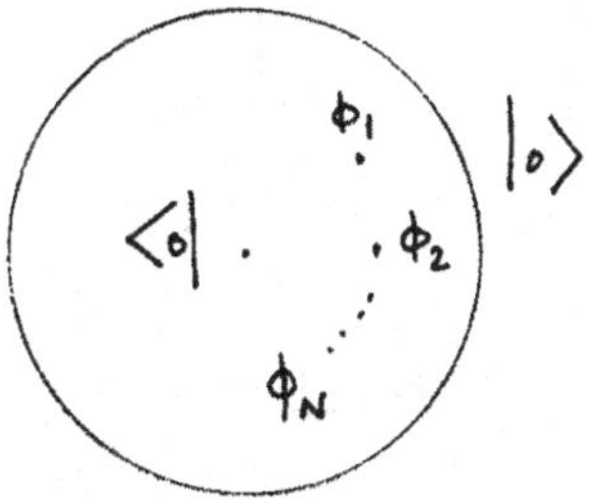

Fig. 10

We want to replace the "in" state $|0\rangle$ by some other state $|W\rangle$, characterizing what happens on the rest of the surface, and such that

$$\frac{\langle 0 | \phi_1 \cdots \phi_N | W \rangle}{\langle 0 | W \rangle} \tag{3.2}$$

agrees with the standard functional integral. Thus if $\mathcal{H}$ is the Hilbert space of the theory, and we consider $\mathcal{P}(g,n) \equiv \{$moduli space of genus g surfaces with n parametrized boundaries, i.e., n points P_i, $i = 1, \ldots, n$ and n local parameters $z_i(P_i) = 0\}$, we want to construct a map

$$\mathcal{P}(g,n) \longrightarrow \mathcal{H}^{\otimes n} \tag{3.3}$$

for each conformal theory. In other words, for every conformal theory, we want to associate a ray with every point $P \in \mathcal{P}(g,n)$, $P = \{X_g, P_1, \ldots, P_n \quad X_g, z_i(P_i) = 0\}$.

Furthermore, we would like to have some consistency conditions enumerated below. In the collection of spaces parametrized by $\mathcal{P}(g,n)$ $\forall$ g,n we have some kind of semi-group structure:

$$\mathcal{P}(g_1,n_1) \times \mathcal{P}(g_2,n_2) \longrightarrow \mathcal{P}(g_1+g_2,\, n_1+n_2-2)$$

$$P_i\infty_j Q = R \qquad P \in \mathcal{P}(g_1,n_1),\; Q \in \mathcal{P}(g_2,n_2) \tag{3.4}$$

Choose two points, i in P and j in Q, and defining R by $z_i z_j = 1$, or by sticking a plumbing fixture. Similarly we have the map $\mathcal{P}(g_1,n_1) \to \mathcal{P}(g_1+1,n_1-2)$, where we glue two points of the same surface. Notice that from this point of view $\mathcal{P}(g,n)$ is generated from $\mathcal{P}(0,3)$ and $\mathcal{P}(0,2)$. In pictures:

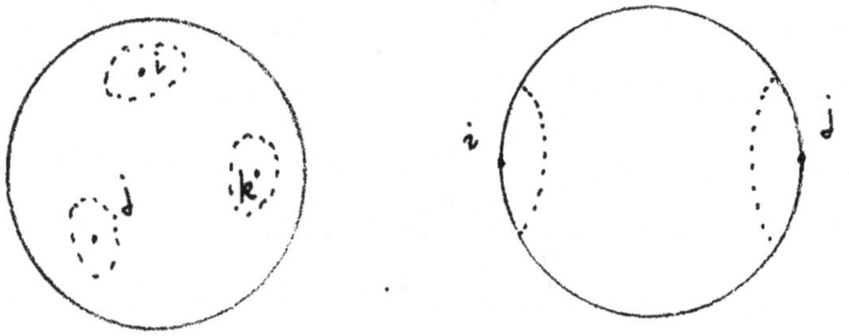

Fig. 11

It is also clear that Fig. 11b acts like a "unit" with respect to sewing (gluing). A representation of these objects for conformal theories contains the information about the OPE C_{ijk} (the three-point function) and the spectrum of the theory, the two-point functions. Let $\{|n\rangle\}$ be an orthonormal basis of states in $\mathcal{H}$. Then one associates to Fig. 11b the state

$$|S_{ij}\rangle = \sum_n |n\rangle_i |n_j\rangle \tag{3.5}$$

(we need a real basis, CPT invariant). Then the state associated to $R = P_i\infty_j Q$ should be

$$|R\rangle = \langle S_{ij}|P\rangle \otimes |Q\rangle \tag{3.6}$$

For theories defined via functional integrals, this is a simple consequence of the properties of the Feynman integral.

Comments

1) Notice that rays are associated to points in $\mathcal{P}(g,n)$ only for conformal theories. Only the normalization of the ray may depend on the metric chosen to represent the conformally invariant data parametrized by $P \in \mathcal{P}(g,n)$. Thus, if we wanted to associated a state to P rather than a ray, we would have to give more information.

2) $\mathcal{P}(g,n)$ is a space more natural than $\mathcal{M}(g,n) = \{$moduli space of genus g curves with n distinguished points$\}$ for several reasons. First, to define oscillators, creation and annihilation operators, etc., in the neighbourhood of a point P, we need a local parameter (and some trivializing sections) to define the Laurent expansion of various fields. Second, if $T_{g,n} \equiv$ mapping class group of X_g with n marked points, then $\pi_1 \mathcal{P}(g,n) = T_{g,n}$. In other words, $\mathcal{P}(g,n)$ resolves the orbifold singularities of $\mathcal{M}_{g,n}$. This is because together with a point $P \in X$ we give a tangent vector at P. Thus, even though X may have conformal automorphisms fixing P (and the holomorphic structure), the tangent vector will change. In fact, the orbifold points always correspond to curves with automorphisms. For instance, the moduli space $\mathcal{M}_1$ (genus 1) is the shaded region in Fig. 12. The points $\tau = i$ and $\tau = \rho$, $\rho^3 = 1$ correspond to the

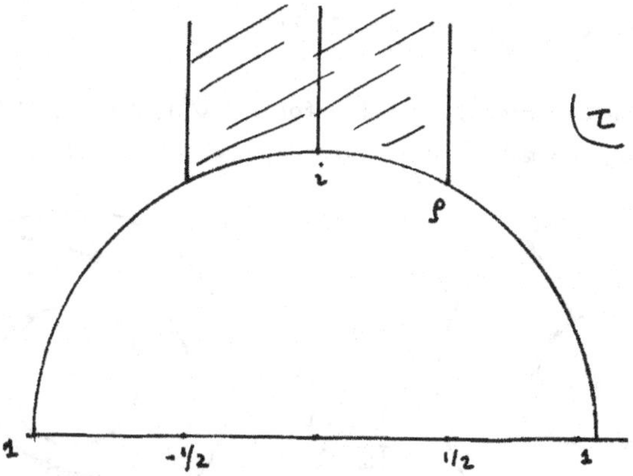

Fig. 12

square torus with a Z_4 symmetry group, and to a unit cell in the SU(3) root lattice respectively. For higher genus surfaces the same conclusion holds, although the automorphism groups are more complicated. In fact, they are all invariant subgroups of the Hurwicz group with presentation

$$H = \langle E_1, E_2 \mid E_1^2 = E_2^2 = (E_1 E_2)^7 \rangle$$

For instance, the famous Klein subgroup of SU(3) of order 168 appears first in genus 3. The modular group $T_{g,n}$ admits a presentation in terms of elements of finite order. This means that we can represent $T_{g,n}$ in terms of the curves left fixed by these generators.

3) Harer has shown that for $g > 2$, $H_1(\mathcal{P}(g,n)) = 0$. This is very useful in the characterization of the Polyakov measure, because it implies that any flat line bundle is necessarily trivial.

4) As we will see in more detail later, the "equations of motion" are given by the energy-momentum tensor. Let $v = v(z)(d/dz)$ [in $\mathcal{P}(g,1)$ for simplicity] represent a variation of the moduli in $\mathcal{P}(g,1)$ at P. Then the change in $|W\rangle$ as we change the moduli is given by

$$\delta_v |W\rangle = \left(T(v) + \overline{T(v)}\right) |W\rangle \tag{3.7}$$

$$T(v) = \oint_P v(z)\, T(z) \tag{3.8}$$

We will prove this in some cases, but it can be taken as an axiom in the construction of CFT for genus > 1. This is essentially the approach taken by Segal in his axiomatic approach to CFT [21].

5) Consistency of the construction of $|W\rangle$ for $P \in \mathcal{P}(g,n)$ requires that $L_0 - \bar{L}_0 \in Z$. This is because we can make a Dehn twist around a point P.

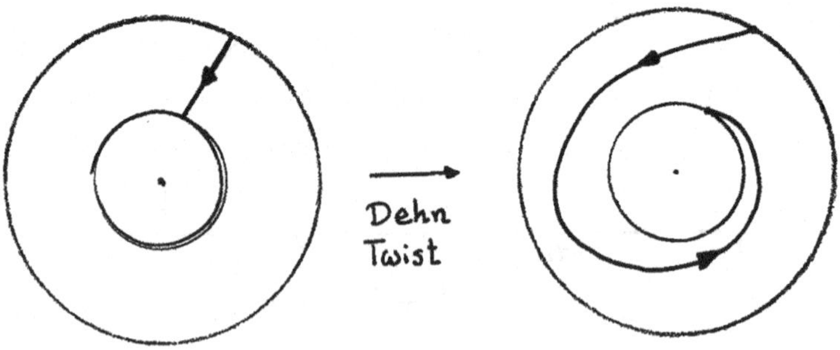

Fig. 13

On the holomorphic co-ordinate z, this amounts to $z \to e^{2\pi i\theta}z$, $0 < \theta \leq 1$. On the state $|W\rangle$, this transformation is implemented by $e^{2\pi i\theta(L_0-\bar{L}_0)}$. When $\theta = 1$, then invariance requires $L_0-\bar{L}_0$ = integer. This is a necessary condition for modular invariance.

Let us now consider some examples to illustrate the general framework.

Free fermions: $\psi(z)$, $\bar{\psi}(z)$, $S = \int \bar{\psi}\bar{\partial}\psi$

In genus g there are 2^{2g} spin structures, $2^{g-1}(2^g-1)$ odd [dimKer $\bar{\partial}$ = 1 (mod 2)] and $2^{g-1}(2^g+1)$ even [dim Ker $\bar{\partial}$ = 0 (mod 2)]. For the time being, choose an even non-singular spin structure for $\psi, \bar{\psi}$, i.e., dim Ker $\bar{\partial} = 0$, $\psi = c(z)$, $\bar{\psi} = b(z)$.

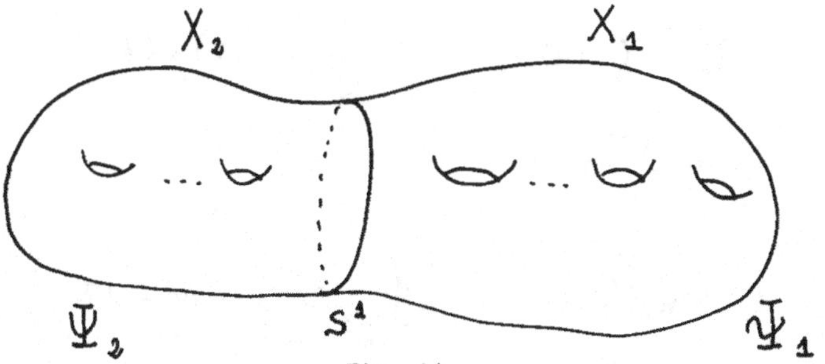

Fig. 14

We want to associate a state to each side of the surface in Fig. 14 so that on the Fock space of the circle

$$Z = \int (db\,dc) \exp - \int b\bar{\partial}c = \langle \Psi_2 | \Psi_1 \rangle \tag{3.9}$$

We can represent $\bar{\Psi}_1$ as follows:

$$\Psi_1[f] = \int_{c|_{S^1} = f} db\,dc \exp\left(- \int_{X_1} c\bar{\partial}b + \oint_{S^1} cb\right) \tag{3.10}$$

b and c behave as holomorphic $\frac{1}{2}$ differentials under conformal transformations $\{c_n, b_m\} = \delta_{n+m,0}$, i.e., $b(z) \sim \delta/(\delta c(z))$; that is, b is the "translation" operator for c.

Let w_n be the boundary value at S^1 of a holomorphic spinor on X_1. Then we want to show that

$$\Psi_1[f + w_n] = \Psi_1[f] \tag{3.11}$$

This simply follows from the definition of $\Psi_1[f]$ and making a change of variables $c \to c - w_n$. Since b is the translation operator for c(z), we can write this equation by saying that

$$Q(w_n) \equiv \oint_{S^1} b(z) w_n(z) \quad ; \quad Q(w_n)\Psi_1 = 0 \tag{3.12}$$

Furthermore, since $w_n(z)$ is holomorphic, if $b(z)$ is also holomorphic, $Q(w_n)$ only depends on the homology class of the contour. This follows because in correlation functions $\bar{\partial} b = 0$, as follows by making a change $c \to c+h$, such that $h|_{S^1} = 0$. This implies that $\bar{\partial} b = 0$ inside correlation functions. We can do similarly with $\tilde{w}_n(z)$, to show that

$$\begin{aligned} \tilde{Q}_n = Q(\tilde{w}_n) &= \oint \tilde{w}_n(z)\, c(z) \\ \tilde{Q}_n |\Psi_1\rangle &= 0 \end{aligned} \tag{3.13}$$

and also

$$\{ Q_n, \tilde{Q}_m \} = \oint_{S^1} w_n(z)\, \tilde{w}_m(z) \tag{3.14}$$

Since w_n, $\tilde{w}_m$ are holomorphic on S^1, $w_n\tilde{w}_m$ is a holomorphic (1,0) form on X_1, and a simple contour deformation argument shows that

$$\{ Q_n, \tilde{Q}_m \} = 0 \tag{3.15}$$

In fact, Q_n, $\tilde{Q}_m$ give the maximal set of conditions on $|\Psi_1\rangle$. For example, for $g = 0$ and one point P, $z(p) = 0$, the holomorphic section in $S^2-\{P\}$ with poles at P has the form: $z^{-n}\, dz^{1/2}$

$$b(z) = \sum_{n \in \mathbb{Z}+1/2} b_n z^{-n-1/2} = \sum_{n \in \mathbb{Z}} b_{n-1/2}\, z^{n-1} \tag{3.16}$$

Then

$$\oint_{S^1} b(z)\, z^{-n} dz = b_{n-1/2} \qquad n \geq 1 \tag{3.17}$$

hence

$$b_n|0\rangle = c_n|0\rangle = 0 \qquad n > 0 \tag{3.18}$$

and we obtain the SL_2-invariant vacuum.

The state can be constructed explicitly using prime forms and θ-functions. Since we have a local parameter, let S(z,w) be the Szegö kernel for spin ½ with the given spin structure (the two-point function for the fermion):

$$S(z,w) = \frac{\vartheta\left[\begin{smallmatrix}\alpha\\ \beta\end{smallmatrix}\right](z-w\,|\,\Omega)}{\vartheta\left[\begin{smallmatrix}\alpha\\ \beta\end{smallmatrix}\right](0|\Omega)\; E(z,w)} \tag{3.19}$$

E(z,w) is the prime for on X. It is the unique $(-\frac{1}{2},-\frac{1}{2})$ differential such that

a) $E(z,w) = -E(w,z)$

b) Its only zero is first order, and it happens along the diagonal: $E(z,w) \sim (z-w) + O((z-w)^3)$ as $z \to w$.

To construct it, let $\left[\begin{smallmatrix}a\\ b\end{smallmatrix}\right]$ be a non-singular, odd theta characteristic. Then:

$$h^2(z) = \sum_{j=1}^{g} \frac{\partial}{\partial u^j}\, \vartheta\left[\begin{smallmatrix}a\\ b\end{smallmatrix}\right]\left(\int_w^z \omega \,\Big|\, \Omega\right)\Big|_{z=w} \omega_j(z) \tag{3.20}$$

has double zeroes at g-1 points $P_1, \ldots, P_{g-1}$ as a consequence of the Riemann vanishing theorem. Then

$$E(z,w) = \frac{\vartheta\left[\begin{smallmatrix}a\\ b\end{smallmatrix}\right](z-w\,|\,\Omega)}{h(z)\; h(w)} \tag{3.21}$$

h(z) is the holomorphic section of $\left[\begin{smallmatrix}a\\ b\end{smallmatrix}\right]$, and the square root in (3.20) does not introduce cuts, because all its zeroes are second order. For z,w in the neighbourhood of $P \in X$, we can write

$$w_n(z) = \frac{1}{(n-1)!} \frac{\partial^{n-1}}{\partial w^{n-1}} S(z,w)\Big|_{w=0}$$

$$= \frac{1}{z^n} + \sum_{m=1}^{\infty} B_{nm}\, z^{m-1}$$

$$B_{nm} = \frac{1}{(n-1)!\,(m-1)!} \frac{\partial^{m-1}}{\partial z^{m-1}} \frac{\partial^{n-1}}{\partial w^{n-1}} \left(S(z,w) - \frac{1}{z-w} \right) \tag{3.22}$$

and now we can write down a differential equation for $|\Psi\rangle$. Since $\{c_n, b_m\} = \delta_{n+m,0}$, $n,m \quad Z + \frac{1}{2}$, we represent the operators as

$$c_n = \frac{\partial}{\partial b_{-n}} \qquad b_n = \frac{\partial}{\partial c_{-n}} \qquad n > 0$$

$$c_{-n} = c_{-n} \cdot \qquad b_{-n} = b_{-n} \cdot \qquad n > 0 \tag{3.23}$$

Hence $\Psi = \Psi(c_{-n}, b_{-m})$, and $|0\rangle \to \Psi = 1$. The conserved charges are

$$\oint_{S^1} b(z)\, w_n(z) = \oint_{S^1} \sum_k b_{k+1/2}\, z^k \left(z^{-n} + \sum_{m=1}^{\infty} B_{nm}\, z^{m-1} \right)$$

$$= b_{n-1/2} + \sum_{m=1}^{\infty} B_{nm}\, b_{-m+1/2}$$

$$= \partial/\partial c_{n+1/2} + \sum_{m \geq 1} B_{nm}\, b_{-m+1/2} \tag{3.24}$$

hence

$$|\Psi\rangle = C \exp\left(- \sum_{n,m=1}^{\infty} B_{nm}\, c_{n+1/2}\, b_{m+1/2} \right) |0\rangle \tag{3.25}$$

C is the (undetermined) normalization of the ray. $|\Psi\rangle$ is a Bogoliubov transformation of $|0\rangle$. Next, consider $P \in \mathcal{P}(g,n)$. Now we have to consider holomorphic sections of $K^{\frac{1}{2}}$ in $X - \{P_1, \ldots, P_n\}$ with poles only at $P_1, \ldots, P_n$. In particular, we can construct $|S_{12}\rangle$:

$$z^n\, dz^{1/2} \quad , \quad n \in \mathbb{Z}$$

$$\mathcal{Q}(w_n) = \oint_{P_1} b^{(1)}(z_1)\, w_n(z_1) + \oint_{P_2} b^{(2)}(z_2)\, w_n(z_2) \tag{3.26}$$

In this case,

$$|S_{12}\rangle = \Big(\prod_{m=1}^{\infty} e^{\,c^{(1)}_{m+1/2}\, b^{(2)}_{-m+1/2} + c^{(2)}_{-m+1/2}\, b^{(1)}_{-m+1/2}}\Big)\, |0\rangle_1\, |0\rangle_2 \tag{3.27}$$

and

$$(\mathcal{Q}_1 + \mathcal{Q}_2)\, |S_{12}\rangle = 0 \tag{3.28}$$

For illustrative purposes, we now prove the charge transport argument (CTA) which implies the sewing rules for spin $\frac{1}{2}$. Let $|P\rangle$ be the state associated to $P \in \mathcal{P}(g_1,n_1)$ and $|Q\rangle$ the one associated to $Q \in \mathcal{P}(g_2,n_2)$. If we sew along the i,j punctures we obtain $R = P_i \infty_j Q \in \mathcal{P}(g_1+g_2,\ n_1+n_2-2)$. We wish to prove that

$$|R\rangle = \underset{z_i z_j = 1}{\langle S_{ij}|P\rangle_i} \otimes |Q\rangle_j \in \mathcal{H}^{\otimes(n_1+n_2-2)} \tag{3.29}$$

Let $P_1, \ldots, P_{i-1}, P_{j+1}, \ldots, P_{n_1+n_2}$ be the distinguished points in R. The conserved charge condition is

$$(\mathcal{Q}_1 + \cdots + \mathcal{Q}_{i-1} + \mathcal{Q}_{j+1} + \cdots + \mathcal{Q}_{n_1+n_2})\, |R\rangle = 0 \tag{3.30}$$

providing the maximal set of conditions on the state $|R\rangle$. To prove that $\langle S_{ij}|P\rangle \otimes |Q\rangle$ satisfies these equations, we use

$$\langle S_{ij}|\, (\mathcal{Q}_i + \mathcal{Q}_j) \tag{3.31}$$

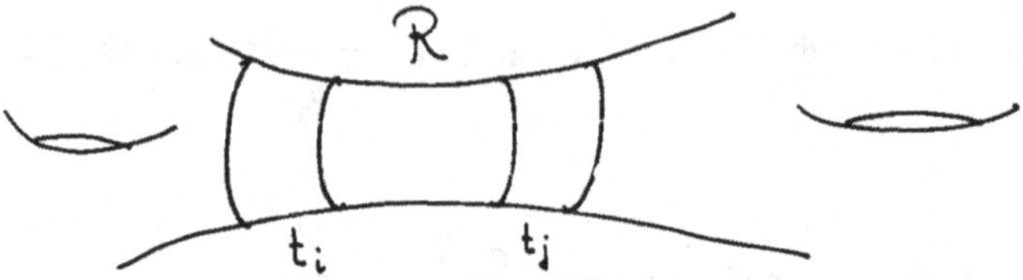

Fig. 15

In R we still have some remnants of the two discs around punctures i (in P) and j (in Q). If we expand any of the holomorphic spinors in R away from $P_1, \ldots, P_{i-1}, P_{j+1}, \ldots$, we can express them as functions of t_i, t_j. As such, they will have at most poles in the (non-existent) points $t_i = t_j = 0$. Therefore, we now have for these spinors $\langle S_{ij}|(Q_i+Q_j) = 0_j$ and

$$(Q_1+\cdots+Q_{i-1}+Q_{j+1}+\cdots+Q_{n_1+n_2})\,\langle S_{ij}|P\rangle\otimes|Q\rangle$$
$$=\langle S_{ij}|\,(Q_1+\cdots+Q_i)+(Q_j+Q_{j+1}+\cdots+Q_{n_1+n_2})\,|P\rangle\otimes|Q\rangle=0$$

by the definition of $|P\rangle$ and $|Q\rangle$. Therefore the sewing condition works. One can similarly construct the state for the sphere with three points. This is left as an exercise to the reader.

LECTURE 4

We continue with the construction of states associated to Riemann surfaces. Now we would like to treat the case of a single-valued scalar field ϕ. This is just one component of the bosonic string moving in flat space. We can start by defining the state from the functional integral point of view on the surface in Fig. 16.

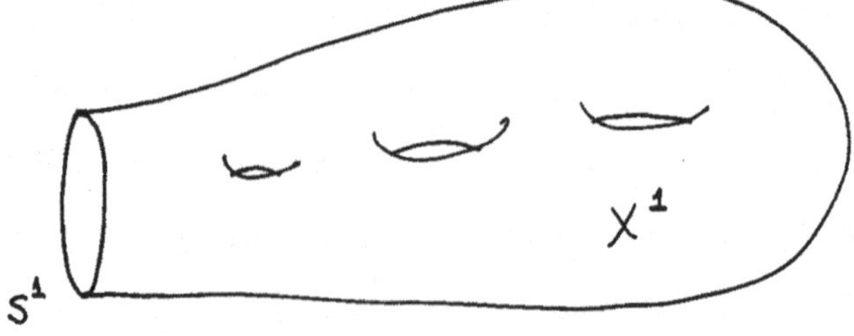

Fig. 16

The state is a functional:

$$\Psi[f] = \int_{\phi_{S^1} = f} d\phi \, \exp\left(-\frac{i}{2}\int_{X^1} \partial\phi\bar{\partial}\phi\right) \tag{4.1}$$

Since f is a real function, we can always construct a harmonic function on X_1 whose boundary value is f. We also call f this harmonic function. Then the functional integral can be computed explicitly. Writing $\phi = f+\phi'$, with $\phi'|_{S^1} = 0$, we have, with * being the duality operator,

$$\begin{aligned} S &= \frac{1}{2}\int d\phi \wedge * d\phi = \frac{1}{2}\int d(f+\phi')\wedge * d(f+\phi') \\ &= \frac{1}{2}\int df\wedge * df + \frac{1}{2}\int d\phi' \wedge * d\phi' + \int d\phi' \wedge * df \end{aligned} \tag{4.2}$$

Since $d*df = 0$ (harmonic function), we obtain

$$S[f+\phi'] = S[\phi] + \frac{1}{2}\oint_{S^1} f * df \tag{4.3}$$

hence

$$\Psi[f] = \left(\det{}'\Delta\right)^{-1/2}_{\phi'|_{S^1}=0} \, e^{-\frac{1}{2}\int_{S^1} f*df} \tag{4.4}$$

The conservation laws are easy to write down. We only need to use Green's theorem

$$Q(h_n) = \oint_{S^1}\left(h_n * d\phi - \phi * dh_n\right) \tag{4.5}$$

where h_n is a harmonic function analytic on X_1. Since in correlation functions the field ϕ satisfies $\partial\bar{\partial}\phi = 0$, it is easy to see that as an operator, $Q(h_n)$ only depends on the homology class of S^1, and not on its detailed shape. Writing locally $h = h_H + h_A$, $\bar{\partial}h_H = \partial h_A = 0$ (holomorphic and antiholomorphic pieces resp.), we can write the conserved current as

$$j(h_n) = \partial\phi \, h_n^H - \bar{\partial}\phi \, h_n^A \tag{4.6}$$

Here we meet a problem in trying to construct the h_n's explicitly. For a generic point P, we can construct functions holomorphic off P with poles of order higher than g at P. In general, between 1 and 2g, there are g gaps in the possible order of the pole at P; this is the Weierstrass gap theorem [12]. Hence we get "∞-g" conditions on the state Ψ. However, we do not have to take purely holomorphic objects, i.e., we can consider holomorphic multivalued functions, and compensate the multivaluedness using antiholomorphic quantities. Explicitly, consider the differentials:

$$\eta_n(t) = \frac{1}{(n-1)!} \frac{\partial}{\partial t} \frac{\partial^n}{\partial y^n} \ln E(t,y) \Big|_{y=0} \tag{4.7}$$

then define

$$h_n(t) = \int^t \left[\eta_n(t) - A_n (\mathrm{Im}\,\Omega)^{-1} (\omega - \bar{\omega}) \right] \tag{4.8}$$

where w_i are the Abelian differentials, and

$$\omega_i(t) = \sum_{n=1}^{\infty} A_{ni}\, t^{n-1} dt \tag{4.9}$$

Moreover, we need to use the property of $\eta_n(t)$ that

$$\oint_{b_i} \eta_n(t) = A_{ni} \tag{4.10}$$

It is easy to check, using the properties of the prime form, that $h_n(t)$ is single-valued. For later purposes, define

$$\ell_{nm} = \frac{1}{2(n-1)!\,(m-1)!} \frac{\partial^n}{\partial t^n} \frac{\partial^m}{\partial y^m} \ln \frac{E(t,y)}{t-y} \tag{4.11}$$

and expand $\phi(t)$ in oscillators:

$$\phi(t) = q + ip \ln t + i \sum_{n \neq 0} \left(\frac{t^n a_n}{n} + c.c. \right) \tag{4.12}$$

We can realize the a_n's by

$$a_n = \frac{\partial}{\partial x_n} \qquad a_{-n} = n x_n \tag{4.13}$$

Writing

$$\mathcal{Q}(h_n)|\Psi\rangle = \bar{\mathcal{Q}}(\bar{h}_n)|\Psi\rangle = 0 \qquad (4.14)$$

as a differential equation, we obtain

$$\langle x,\bar{x}|\Psi\rangle = \exp\ (x,\bar{x})\, M \begin{pmatrix} x \\ \bar{x} \end{pmatrix}$$

$$M = \begin{pmatrix} \mathcal{Q}_{nm} + \frac{\pi}{2} A_n \Omega_2^{-1} A_m & -\frac{\pi}{2} A_n \Omega_2^{-1} \bar{A}_m \\ -\frac{\pi}{2} \bar{A}_n \Omega_2^{-1} A_m & \bar{\mathcal{Q}}_{nm} + \frac{\pi}{2} \bar{A}_n \Omega_2^{-1} \bar{A}_m \end{pmatrix} \qquad (4.15)$$

If we want to consider $|S_{12}\rangle$, then

$$|S_{12}\rangle = \int dp \prod_{n=1}^{\infty} \exp\left(a_n^{(1)+} a_m^{(2)+} + \bar{a}_n^{(1)+} \bar{a}_n^{(2)+}\right) |p\rangle_1 \otimes |-p\rangle_2 \qquad (4.16)$$

and finally, for an arbitrary number of points, let

$$G(z,w) = -2\pi \left(\operatorname{Im}\int_z^w \omega\right)(\operatorname{Im}\Omega)^{-1}\left(\operatorname{Im}\int_z^w \omega\right) - \ln|E(z,w)|^2 \qquad (4.17)$$

If z_i, w_j coincide on the same patch, then define

$$g(z_i,w_j) = G(z_i,w_j) + \ln|z_i - w_j|^2 \qquad (4.18)$$

otherwise $g = G(z_i,w_j)$. Then, for n punctures,

$$|\Psi_g^n\rangle = \int dp_1 \ldots dp_n\, \delta(\Sigma p_i) \exp\left\{ -\frac{1}{(2\pi i)^2} \sum_{i,j} \oint_i \oint_j (\partial x^{(i)}, \bar{\partial} x^{(i)})\, M(z_i,z_j) \begin{pmatrix} \partial x^{(j)} \\ \bar{\partial} x^{(j)} \end{pmatrix} \right\} |p_1\rangle \ldots |p_n\rangle \qquad (4.19)$$

where $M(z_i,z_j)$ is constructed out of $g(z_i,z_j)$ in an obvious way. Once again, it is not difficult to prove the CTA argument for the conserved charges.

Comments

1) For arbitrary conformal theories, if there is a conserved current of spin j, $\bar{\partial} j(z) = 0$, we can construct conserved charges in the same way:

$$Q_n = \oint_{S^1} j(z)\, w_n(z)$$

where $w_n(z)$ are (1-j)-differential holomorphic off P.

2) The construction can easily be extended to multivalued fields, and in this way we can describe scalar fields ϕ^i taking values on some d-dimensional torus $\mathbb{R}^d/\Lambda^d$ (Λ^d being a d-dimensional lattice).

3) Combining the results we have presented with remark 2), the computations of the last section and some amount of information on the Hilbert-Schmidt Grassmannian, it is possible to prove the bosonization formulae for higher genus Riemann surfaces. In particular, one of the consequences of the proof is Fay's trisecant identity, which plays a central rôle in the K-P hierarchy and the proof of the Novikov conjecture.

The ghost system (spin j = 2):

Consider first the sphere with one singled-out point P, z(p) = 0. Since

$$b(z) = \sum_{n \in \mathbb{Z}} b_n z^{n-2} \qquad c(z) = \sum_n c_n z^{n+1}$$

$$\{b_n, c_m\} = \delta_{n+m,0} \tag{4.20}$$

we can construct the vector fields holomorphic off P. They are

$$\{z^2, z, 1, z^{-1}, z^{-2}, \ldots\} \tag{4.21}$$

For quadratic differentials, those extending holomorphically off P are

$$\{z^{-4}, z^{-5}, z^{-6}, \ldots\} \tag{4.22}$$

These results follow from the Riemann-Roch theorem on the sphere [12]. The conserved charges are then

$$k = 2,1,0,-1,-2,\ldots \qquad \oint b_m z^{m-2} z^k = b_{-k+1} \quad \text{i.e. } b_{-1}, b_0, b_1, \ldots$$

$$k = 4,5,\ldots \qquad \oint c_m z^{m+1} z^{-k} = c_{k-2} \quad \text{i.e. } c_2, c_3, c_4 \ldots \tag{4.23}$$

Hence the state $|\phi_o\rangle$ is defined by

$$\begin{aligned} b_n |\phi_0\rangle &= 0 \qquad n > -2 \\ c_n |\phi_0\rangle &= 0 \qquad n > 1 \end{aligned} \tag{4.24}$$

It is easy to show that this state is $SL_2(C)$-invariant, i.e., $L_o|\phi_o\rangle = L_{\pm 1}|\phi_o\rangle = 0$. Its charge assignment is a bit more delicate. We have a ghost number current $j(z) = :c(z)b(z):$, and we have to determine the vacuum charge. Since

$$\begin{aligned} [L_0, b_n] &= -n b_n \\ [L_0, c_n] &= -n c_n \end{aligned} \tag{4.25}$$

$|\phi_o\rangle$ is not a highest weight of the oscillator algebra (i.e., there are negative energy modes which do not annihilate $|\phi_o\rangle$). Similarly, we have a zero mode algebra $\{c_o, b_o\} = 1$. We introduce two highest-weight states $|\pm\rangle$:

$$\begin{aligned} c_n|+\rangle &= b_n|-\rangle = 0 \qquad n \geq 0 \\ c_0|-\rangle &= |+\rangle \qquad b_0|+\rangle = |-\rangle \end{aligned} \tag{4.26}$$

It is natural to assign $Q|\pm\rangle = \pm\frac{1}{2}|\pm\rangle$, $Q = \int_P j(z)$. These states are obtained from $|\phi_o\rangle$ as follows:

$$|-\rangle = c_1|\phi_0\rangle \qquad |+\rangle = c_0 c_1 |\phi_0\rangle$$

$$b_0|+\rangle = b_0 c_0 c_1 |\phi_0\rangle = c_1|\phi_0\rangle - c_0 b_0 c_1 |\phi_0\rangle = c_1|\phi_0\rangle \tag{4.27}$$

Then

$$L_0|\pm\rangle = -1|\pm\rangle \tag{4.28}$$

(This is the origin of the tachyon, as we will see.) A similar construction can be carried out for arbitrary spin, and it is left as an exercise. The next thing we need is the sewing state. On a sphere with two points, the quadratic differentials and vectors are holomorphic off 0, and ∞ are

$$z^n dz^2 \qquad z^{n+1}\frac{d}{dz} \qquad \forall n \in \mathbb{Z} \tag{4.29}$$

Now we have two Hilbert spaces, one for each puncture. The conserved charges are

$$\left(b_n^{(1)} - b_{-n}^{(2)} \right) | S_{12} \rangle = 0 \qquad n \in \mathbb{Z} \tag{4.30}$$

and similarly for the c's. Then

$$| S_{12} \rangle = \prod_{m=1}^{\infty} \exp\left(- c_{-m}^{(1)} b_{-m}^{(2)} - c_{-m}^{(2)} b_{-m}^{(1)} \right) \left(b_0^{(1)} - b_0^{(2)} \right) |+\rangle_1 |+\rangle_2 \tag{4.31}$$

The total charge is

$$\mathcal{Q} \, | S_{12} \rangle = \left(Q^{(1)} + Q^{(2)} \right) | S_{12} \rangle = 0 \tag{4.32}$$

For higher genus surfaces, let $|\phi_g^n\rangle$ be the state corresponding to a genus g surfaces with n distinguished points. First, we know that for $g > 1$, there are 3g-3 holomorphic quadratic differentials. Similarly, we can construct at every point $P \in X$ quadratic differentials with arbitrary order poles. At a particular point P with local parameter z, we can write the holomorphic quadratic differentials as

$$\psi_{n+1} = z^n + \sum_{m \geq q} C_{nm}^{(2)} z^m \qquad q = 3g-3$$
$$n = 0, 1, \ldots, 3g-3 \tag{4.33}$$

and those with poles as

$$S_n = z^{-n} + \sum_{m \geq q} B_{nm}^{(2)} z^m \tag{4.34}$$

and similarly for vectors v_n. The coefficients B can be written in terms of prime forms and θ-functions. Note that

$$\oint c(z)\, \psi_{n+1}(z) = \oint \sum_k c_k z^{k+1} \left(z^n + \sum_{m \geq q} C_{nm}^{(2)} z^m \right) =$$
$$= c_{-n-2} + \sum_{m \geq q} C_{nm}^{(2)} c_{-m-2} \tag{4.35}$$

i.e., it only involves creation operators. In the same way,

$$\oint c(z)\, s_1(z) = c_{-1} + \sum_{m \geq q} B^{(2)}_{1m}\, c_{-m-2} \qquad (4.36)$$

hence the only way the state $|\phi^n_g>$ can be annihilated by these charges is if they appear explicitly in the state (due to the fermionic character of b and c). For quadratic differentials with higher-order poles, we get both creation and annihilation operators. In this way, we get a differential equation if we represent

$$\{ c_n, b_m \} = \delta_{n+m,0} \qquad (4.37)$$

as

$$c_n = \frac{\partial}{\partial b_{-n}} \quad n \geq 0 \qquad b_n = \frac{\partial}{\partial c_{-n}} \quad n > 0 \qquad (4.38)$$

so that the wave functionals become functions of b_{-n}, $n > 0$, c_{-n}, $n > 1$, and $|+> = 1$. Hence, the state $|\phi^n_g>$ takes the form:

$$|\phi^n_g> = c_1 \ldots c_{3g-3}\, A^{(1)}_1 \ldots A^{(n)}_1 \exp - \sum_{\substack{m \geq q \\ n \geq 2}} B^{(2)ij}_{nm}\, c^{(i)}_{-m-2}\, b^{(j)}_{-n+2} \; |+>_1 \otimes \ldots \otimes |+>_n \qquad (4.39)$$

where

$$A^{(i)}_1 = \sum_j \oint_{P_j} s^{(i)}_{-1}(z_j)\, c^{(j)}(z_j) \qquad (4.40)$$

and $s^{(i)}_{-1}(z)$ is the quadratic differential with a single pole at P_i. Finally,

$$c_{i+1} = \sum_j \oint_j \psi_i(z_j)\, c^{(j)}(z_j) \qquad (4.41)$$

The ghost charge of $|\phi^n_g>$ is easy to count once we recall that b (c) has ghost number -1 (+1):

$$Q_{gh} |\phi^n_g> = (3g - 3 + 3n/2)\, |\phi^n_g> \qquad (4.42)$$

Notice that this assignment satisfies the sewing rules. If we sew a state in $\mathcal{P}(g_1, n_1)$ and another in $\mathcal{P}(g_2, n_2)$, then their charges add up correctly:

$$[\,3(g_1-1)+3n_1/2\,]+[\,3(g_2-1)+3n_2/2\,]=$$

$$=3(g_1+g_2-1)+3(n_1+n_2-2)/2 \qquad (4.43)$$

Furthermore, the charge transport argument applies here without modifications.

LECTURE 5

Next we would like to show how the Virasoro algebra gives a connection over $\mathcal{P}(g,n)$. In other words, in the space of states $\mathcal{H}^{\otimes n}$, for an arbitrary CFT, we have obtained a ray $|\phi>_P$ for each $\mathcal{P}(g,n)$. We would like to know how to transport the ray $|\phi>_P$ along different curves in $\mathcal{P}(g,n)$. It is clear in the examples discussed that the ray $|\phi>_P$ is only dependent on the conformal data at $P \in \mathcal{P}(g,n)$, hence $|\phi>$ is uniquely defined once P and the CFT are given. This is important, because it immediately shows that $|\phi>$ is invariant under the action of the modular group, leaving invariant the data parametrized by $\mathcal{P}(g,n)$. The action of the Virasoro algebra on the states $|\phi>$ provides an "equation of motion" for the state which can be used to compute chiral determinants. To understand the action of Virasoro (Vir), consider the infinitesimal Kodaira-Spencer construction. Since in our data we have a curve X, a point P and a local parameter, consider the covering of X shown in Fig. 17.

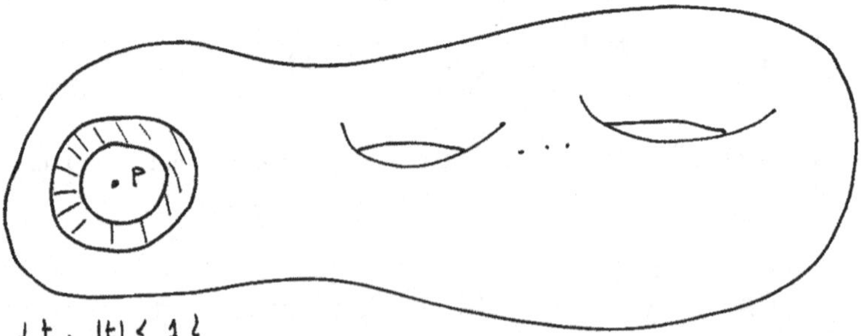

$D = \{t,\ |t| \leq 1\}$
$A = \{r < |t| < 1\}$

Fig. 17

Take $X_1 = X-(|t| < r)$. For a meromorphic vector $v(t)(\partial/\partial t)$, $v(t) = \Sigma a_{n-1}t^n$, $v(t)$ i holomorphic in D-P. For every t on the annulus A, implement the change $t \to t + \epsilon v(t$ (Fig. 18):

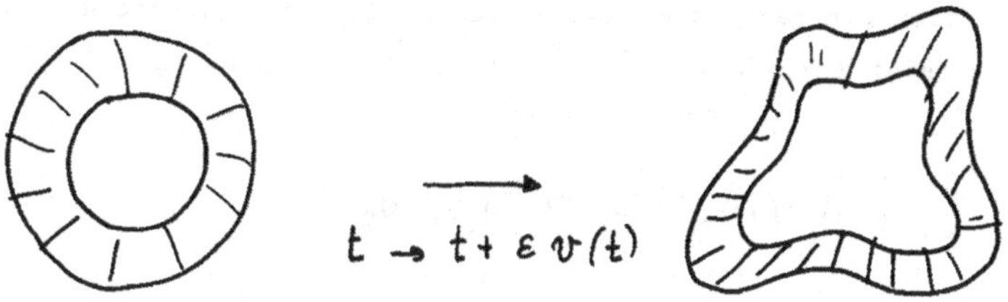

Fig. 18

Then identify the "deformed annulus" point by point with the original one on the curve X. There are now three possibilities:

i) v extends to a holomorphic vector on D and v(o) = 0. This corresponds to a change in the local parameter. If v(o) = 0, this is equivalent to an infinitesimal change in the point P.

ii) v extends holomorphically to X_1. In this case, the data parametrized by $P \in \mathcal{P}(g,1)$ does not change. This vector is an infinitesimal conformal isometry of the rest of the surface.

iii) v does not extend holomorphically to the disc or to the rest of the surface. using Riemann-Roch, the vector fields with this property are:

$$\left\{ \frac{1}{z}, \frac{1}{z^2}, \ldots, \frac{1}{z^{3g-3}} \right\} \tag{5.1}$$

and they represent the tangent space to moduli space $\mathcal{M}_g$ at X. This follows from two facts: (1) $H^1(X,K^{-1})$ is the holomorphic tangent space to $\mathcal{M}_g$ at X; (2) use the sheaf sequence (for P not a Weierstrass point):

$$0 \longrightarrow K^{-1} \longrightarrow K^{-1}\big((3g-3)P\big) \longrightarrow S_P^{3g-3} \longrightarrow 0 \tag{5.2}$$

Here S_P^{3g-3} is the skyscraper sheaf of length 3g-3 at P, $K^{-1}\big((3g-3)P\big)$ is the sheaf of meromorphic vectors with at most order 3g-3 poles at P, and K^{-1} is the sheaf of local holomorphic vectors. When P is generic, the line bundle $K^{-1}\big((3g-3)P\big)$ has degree g-1, and it is away for the Θ-divisor.

Using the long exact sequence and $H^0\big(X,K^{-1}((3g-3)P)\big) = H^1\big(X,K^{-1}((3g-3)P)\big) = 0$, we obtain $H^1(K^{-1}) \simeq S_P^{3g-3} \equiv$ (Laurent tails at P with 3g-3 terms). The Kodaira-

Spencer construction can be exponentiated locally in a neighbourhood of X to provide vector fields passing through $P \in \mathcal{P}(g,1)$: $V(v)$. Those vector fields satisfy the algebra

$$[\, V(v_1), V(v_2) \,] = V(\, [v_1, v_2] \,) \tag{5.3}$$

In a more axiomatic approach to CFT; $|\phi\rangle$ is a well-defined ray at each $P \in \mathcal{P}(g,1)$, and therefore it should be possible to represent the vectors $V(v)$ in terms of operators acting on the Hilbert space . However, since $|\phi\rangle$ is a ray, all we can expect is that the commutation relations of the operator $O(v)$ representing $V(v)$ be:

$$[\, O(v_1), O(v_2) \,] = O(\, [v_1, v_2] \,) + \text{c-number} \tag{5.4}$$

In CFT, the operator $O(v)$ is naturally the energy-momentum tensor. This can be checked for any of the examples of previous lectures, or taken as one of the axioms for CFT's on higher-genus surfaces. Explicitly, the change of $|\phi\rangle$ along v is given by:

$$\delta_v |\phi\rangle = (\, T(v) + \overline{T(v)} \,)\, |\phi\rangle \tag{5.5}$$

This implies that $T(v)$ provides a projective connection over $\mathcal{P}(g,1)$. To check that this holds in the simple examples we analyzed previously, all we need to show is that the new conserved charges are:

$$Q_i \rightarrow Q_i + \epsilon\, [\, T(v), Q_i \,] = Q_i + \delta Q_i \tag{5.6}$$

Then the ray $(1 + \epsilon T(v))|\phi\rangle$ is annihilated by $(Q_i + \delta Q_i)$:

$$\begin{aligned} &(\, Q_i + \delta Q_i)(\, 1 + \epsilon T(v)\,)\, |\phi\rangle = \\ &= \epsilon\, Q_i T(v)\, |\phi\rangle + \epsilon\, (\, T(v)\, Q_i - Q_i T(v)\,)\, |\phi\rangle \\ &= \epsilon\, T(v)\, Q_i\, |\phi\rangle = 0 \end{aligned} \tag{5.7}$$

Following the Kodaira-Spencer construction in detail, the change of a holomorphic section s_i of some holomorphic line bundle with a pole at P is $s_i \rightarrow s_i + \epsilon \mathcal{L}_v s_i$. Since by standard operator algebra

$$[\, T(v), Q(s_i) \,] = Q(\, \mathcal{L}_v s_i \,) \tag{5.8}$$

we conclude that all the states defined in previous sections do satisfy the "equation of motion", i.e., $T(v)$ provides a projective connection.

Comments

1) If we want to construct a state rather than a ray, we need to fix a normalization $C|\phi\rangle$. Imposing the parallel transport equation gives a differential equation for C:

$$\delta_v \log C = \frac{\langle 0| T(v) + c.c.\ |\phi\rangle}{\langle 0|\phi\rangle} \tag{5.9}$$

In particular, for the spin-½ fermion on the torus, the moduli is generated by $v = (1/z)(d/dz)$. Then

$$\delta_v \ln C = B_{12}$$

$$\frac{\partial}{\partial\tau} \ln C = \frac{\partial}{\partial\tau} \ln \frac{\vartheta\left[{\alpha \atop \beta}\right](0|\tau)}{\eta(\tau)} \tag{5.10}$$

In higher genus,

$$\delta_n \ln C = \sum_{m=1}^{n} \left(m - \frac{n+1}{2}\right) B_{n-m+1,\, m} \tag{5.11}$$

where δ_n corresponds to motions along $z^{-n+1}(d/dz)$.

2) Hence, to summarize:

$$\begin{aligned} &L_{-n}\ |\phi\rangle,\ n > 0 && \text{changes local coordinates} \\ &L_{+1}\ |\phi\rangle && \text{moves the point } P \end{aligned} \tag{5.12}$$

$L_k\ |\phi\rangle$, $k = 2,3, \ldots, 3g-3+1$ changes the moduli

3) If v extends holomorphically off P, what should $T(v)|\phi\rangle$ be? Since v does not change any data, all we can expect is that $T(v)|\phi\rangle \propto |\phi\rangle$. One might be tempted to set $T(v)|\phi\rangle = 0$, because $T(z)$ is conserved. However, the anomalous commutator

$$[T(v_1), T(v_2)] = T([v_1,v_2]) + \frac{c}{12}\oint_P v_1 v_2''' \tag{5.13}$$

does not allow this possibility. Another way of looking at the same problem is that $T(z)$ does not transform as a quadratic differential when we move from one patch to another. Under an infinitesimal co-ordinate transformation, $T(v)$ changes by $\int_P \varepsilon'''(z)v(z)$. However, if we choose $\varepsilon'''(z) \in H^0(X,K^2)$, it seems that we obtain $(3g-3)+3$ ways of patching the Riemann surface so that T behaves like a quadratic differential. This is related to the classification of projective structures on a Riemann surface.

The Polyakov Measure

So far, we have associated states to points in $\mathcal{P}(g,n)$. Let us now come back to the bosonic string. As we argued in Lecture 3, we need to construct an integration measure over $\mathcal{M}_g$ or $\mathcal{M}_{g,n}$ (resp. the moduli space of genus g curves, and the moduli space of genus g curves with n marked points). The first one allows the computation of partition functions, and the second one the computation of scattering measures and scattering amplitudes.

It is clear from our construction that there is a lot of redundant information in $\mathcal{P}(g,1)$. There is a projection $\pi(X,P,z) = X$, which simply means forgetting the point and the local parameter.

$$\begin{array}{c} (X,P,z) \in \mathcal{P}(g,1) \\ \big\downarrow \pi \\ \mathcal{M}_g \end{array}$$

Fig. 19

At a point $X \in \mathcal{M}_g$, choose a basis of tangent vectors $V_1, \ldots, V_{3g-3}$ (and c.c.). We want to construct a measure at X:

$$\mu(X)\,(V_1,\ldots,V_{3g-3};\ \overline{V}_1,\ldots,\overline{V}_{3g-3}) \tag{5.14}$$

Given V_i, we can find in $\mathcal{P}(g,1)$ at some point in the fibre $\pi^{-1}(X)$ representatives $v_i(z)\partial/\partial z$ of V_i. The obvious candidate for the measure is

$$\mu(X) = \langle 0|\, b(v_1)\ldots b(v_{3g-3})\,\overline{b(v_1)}\ldots\overline{b(v_{3g-3})}\,|\phi\rangle_{(X,P,z)} \tag{5.15}$$

where

$$b(v) = \oint_P b(z)\, v(z) \tag{5.16}$$

$|0\rangle$ is the SL_2-invariant vacuum in $\mathcal{F}_{matter} \otimes \mathcal{F}_{bc} \otimes \mathcal{F}_{\overline{bc}}$, i.e., $|\phi\rangle$ also belongs to $\mathcal{F}_{matter} \otimes \mathcal{F}_{bc} \otimes \mathcal{F}_{\overline{bc}}$; it is the total state for the free bosonic string. We must show that

i) $\mu(X)$ is independent of the choice of representative of V.

ii) $\mu(X)$ is independent of the local parameter and the position of the point $P \in X$:

iii) $\mu(X)$ satisfies the Belavin-Knizhnik theorem:

$$\partial\bar{\partial} \ln \mu = -13\, \partial\bar{\partial} \ln \operatorname{Im} \Omega \tag{5.17}$$

i.e.,

$$\mu = \frac{\rho \wedge \bar{\rho}}{(\det \operatorname{Im} \Omega)^{13}} \tag{5.18}$$

where ρ is a holomorphic half-volume, the Mumford form.

iv) To prove the last part of iii), we have to show that on surfaces with nodes, ρ has a second-order pole with respect to the holomorphic co-ordinate q such that $q = 0$ defines the curve with a node.

v) If we consider the total energy-momentum tensor

$$T(z) = T_{matter}(z) + T_{gh}(z) \tag{5.19}$$

then $c^{total} = 0$. Hence, if v is a vector field extending holomorphically off P, we can normalize $|\phi\rangle$ so that $T(v)|\phi\rangle = 0$. Since T is the connection acting on $|\phi\rangle$ and defining parallel transport, we conclude that $|\phi\rangle$ is a flat bundle over $\mathcal{P}(g,1)$. This means that locally we can define the normalization of the ray $|\phi\rangle$. It may happen, however, that since $\pi_1(\mathcal{P}(g,1)) = T_{g,1}$ there is non-trivial holonomy for the line bundle $|\phi\rangle$. Since $H_1(\mathcal{P}(g,1)) = 0$, i.e., $T_{g,1}/[T_{g,1}, T_{g,1}] = 0$, a flat line bundle is a trivial line bundle, and the normalization of $|\phi\rangle$ can be defined consistently over all of $\mathcal{P}(g,1)$ up to a constant. This implies among other things that $T(v)|\phi\rangle = 0$ for any v extending holomorphically. If we now change a representative v of moduli change by $v + v_{hol}$ (v_{hol} extends holomorphically off P), then in the measure we have a change

$$\langle 0| \ldots b(v + v_{holo.}) \ldots |\phi\rangle \tag{5.20}$$

Since $b(v_{hol})$ is a conserved charge, and the b's anticommute, then $b(v_{hol})|\phi\rangle = 0$, and the measure does not change. A change in the local parameter is accomplished via L_{-n}, $n > 0$. Since $\langle 0|L_{-n} = 0$, $n > 0$, we conclude, using $[T(v), b(v')] = b([v,v'])$ that there is no dependence on the local parameter. Similarly, since $\langle 0|L_1 = 0$, we can change the point $P \in X$ infinitesimally, and $\mu(X)$ will not vary.

To prove iii), we use the equation of motion and the fact that the matter state does not factorize holomorphically into left and right components. From the equation of motion:

$$\partial\bar{\partial} \log \mu(\ldots V \ldots) = \frac{\langle 0| \ldots T(\,)\,\bar{T}(\,)\,|\phi\rangle}{\langle 0|\phi\rangle} - \frac{\langle 0| \ldots T(\,)\,|\phi\rangle \langle 0| \ldots \bar{T}(\,)\,|\phi\rangle}{\langle 0|\phi\rangle^2} \tag{5.21}$$

Since only the matter sector has factorization problems, we can compute the previous expression in the matter sector. Using Rohrlich's formula

$$\delta \Omega_{ij} = - \oint_P \omega_i \omega_j v \tag{5.22}$$

we obtain

$$\partial\bar{\partial} \log \mu(\ldots V \ldots) = -13\, \partial\bar{\partial} \log \det \operatorname{Im} \Omega \tag{5.23}$$

Finally, to show that the measure at the boundary has second-order poles, we use sewing. Using the plumbing fixture description of degenerating surfaces (Figs. 2 and 21):

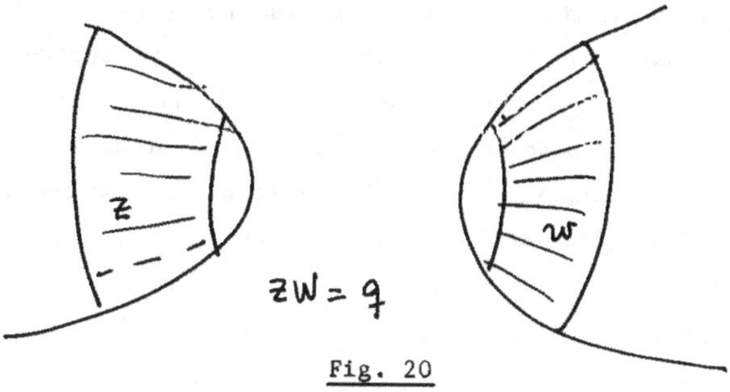

Fig. 20

As $q \to 0$, we generate the curve with a node.

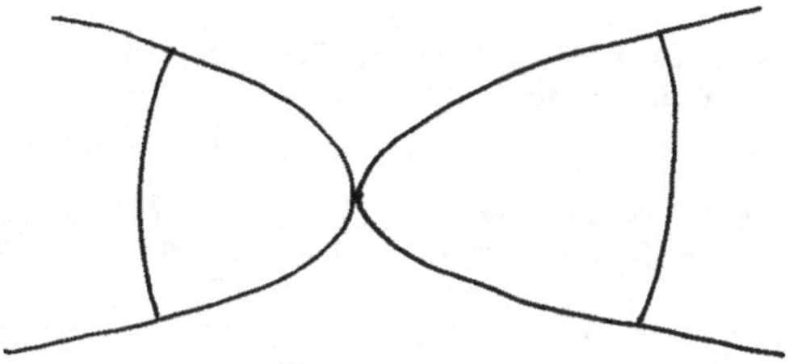

Fig. 21

With the sewing prescription, the measure can be written as

$$\frac{dq\,d\bar{q}}{q\bar{q}} \langle S_{ij} | b_0 \bar{b}_0 \, q^{L_0} \bar{q}^{\bar{L}_0} | \Phi \rangle \tag{5.24}$$

Since the lowest eigenvalue of $L_0, \bar{L}_0$ is -1 (the tachyon), as $q \to 0$ the leading singularity comes from $L_0 = \bar{L}_0 = -1$, and we obtain the behaviour

$$\left| \frac{dq}{q^2} \right|^2 \tag{5.25}$$

Hence, ρ has a second-order pole at the boundary of moduli space, generated by the equivalence classes of curves with one node (either splitting or non-splitting). This property uniquely identifies ρ in μ as the Mumford form, which is the desired result.

Similar arguments can be used to define scattering measures and physical state conditions, and we refer the reader to Ref. [16] for further details. The extension of this construction to fermionic strings is also possible (see the last entry in Ref. [16]), but now we have to adapt the arguments to super-Riemann surfaces. Questions concerning the uniqueness and properties of the supermeasure are answered in a manner similar to the bosonic construction. The problem which still remains unresolved is the correct prescription to integrate over super-moduli space. Details can be found in Ref. [16] (last entry).

REFERENCES

[1] For a thorough account of string theory with many references to the literature, see:
M.B. Green, J.H. Schwarz and E. Witten, "Superstring Theory", vols. I and II, Cambridge University Press (1986).

[2] See, for instance:
J. Kogut and K. Wilson, Physics Reports 12 (1974) 75;
A.M. Polyakov, "Gauge Fields and Strings", Harwood Academic Press, N.Y. (1987);
J. Cardy, Lectures at the 1988 Les Houches Summer School, eds. E. Brezin and J. Zinn-Justin.

[3] D. Friedan, Z. Qiu and S. Shenker, Phys. Rev. Lett. 52 (1984) 1575.

[4] A.A. Belavin, A.M. Polyakov and A. Zamolodchikov, Nucl. Phys. B241 (1984) 333.

[5] D. Friedan, Lectures at the 1983 Les Houches Summer School, eds. J.B. Zuber and R. Stora.

[6] V. Kac, in "Lecture Notes in Physics", Springer Verlag (1978).

[7] V. Dotsenko and V.I. Fateev, Nucl. Phys. B240 (1984) 312; B251 (1985) 691; Phys. Lett. 154B (1985) 291.

[8] E. Verlinde, Utrecht preprint THU-88/17 (1988).

[9] G. Moore and N. Seiberg, Princeton preprint IASSNS-HEP-88/8 (1988).

[10] J. Fröhlich, in Proceedings of the Cargèse Summer School on "Progress in Field Theory and Statistical Mechanics" (1986), eds. A. Jaffe et al., to be published in Plenum Press.

[11] E. D'Hoker and P.H. Phong, Princeton University preprint PUPT-1039 (1988), to appear in Rev. Mod. Phys.

[12] H. Farkas and I. Kra, "Riemann Surfaces", Springer Verlag (1982).

[13] For a readable account of Mumford forms and references to the relevant literature, see:
P. Nelson, "Lectures on Supermanifolds and Strings", to appear in the Proceedings of the Theoretical Advanced Study Institute, Brown University, June 1988.

[14] G. Segal, talk at the Rindberg Schloss Meeting (March 1987), and to appear in Comm. Math. Phys.

[15] A. Neveu and P. West, Phys. Lett. 168B (1985) 192; Nucl. Phys. B278 (1986) 601; Phys. Lett. B193 (1987) 187; Comm. Math. Phys. 114 (1988) 613; Phys. Lett. B194 (1987) 200; B200 (1987) 275; see also:
P. West, "A Review of Duality, String Vertices, Overlap Identities and the Group Theoretical Approach to Strings", CERN preprint TH.4819/87, and references therein.

[16] P. Di Vecchia, R. Nakawaya, J.L. Petersen and S. Sciuto, Nucl. Phys. B282 (1987) 179;
P. Di Vecchia, R. Nakayama, J.L. Petersen, J. Sidenius and S. Sciuto, Phys. Lett. B182 (1986) 164;
P. Di Vecchia, M. Frau, A. Lerda and S. Sciuto, Nucl. Phys. B298 (1988) 526; Phys. Lett. B199 (1987) 49;
P. Di Vecchia, K. Hornfeck, M. Frau, A. Lerda and S. Sciuto, Phys. Lett. B211 (1988) 301;
P. Di Vecchia, F. Pezzella, M. Frau, K. Hornfeck, A. Lerda and S. Sciuto, NORDITA preprint 88/47P (1988).

[17] P. Di Vecchia, R. Nakayama, J.L. Petersen, J. Sidenius and S. Sciuto, Nucl. Phys. B287 (1987) 621;
J.L. Petersen and J. Sidenius, Nucl. Phys. B301 (1988) 247;
J.L. Petersen, K.O. Roland and J.R. Sidenius, Phys. Lett. B205 (1988) 262;
K.O. Roland, Niels Bohr Inst. preprint NBI-HE-88-21;
J.L. Petersen, J.R. Sidenius and A. Tollstén, NBI-HE-88-30.

[18] A. Leclair, Nucl. Phys. B297 (1988) 607; B303 (1988) 189;
A. Leclair, M.E. Peskin and C.R. Preitschopf, SLAC-PUB-4306, 4307 and 4464 (1988).

[19] N. Ishibashi, Y. Matsuo and H. Ooguri, Mod. Phys. Lett. A2 (1987) 119;
L. Alvarez-Gaumé, C. Gomez and C. Reina, Phys. Lett. B190 (1987) 55;
C. Vafa, Phys. Lett. B190 (1987) 47;
L. Alvarez-Gaumé, C. Gomez, G. Moore and C. Vafa, Nucl. Phys. B303 (1988) 455;
L. Alvarez-Gaumé, C. Gomez, P. Nelson, G. Sierra and C. Vafa, CERN preprint TH.5018/88, to appear in Nucl. Phys. B.

[20] D. Friedan and S. Shenker, Phys. Lett. B175 (1986) 287; Nucl. Phys. B281 (1987) 509.

[21] See Ref. [2].

Geometrical aspects of the Kadomtsev–Petviashvili equation

by

Enrico Arbarello and Corrado De Concini

Introduction

In these lectures we focus our attention on the geometrical feature of the theory of the Kadomtsev - Petviashvili (KP) equation. It was discovered by Krichever that the theta function associated to the Jacobian of a Riemann surface satisfies the KP equation.

$$\frac{\partial}{\partial x}(2u_{xxx} - 3uu_x - u_t) + 3u_{yy} = 0.$$

This fact was extensively studied by the school of Novikov, and it is due to Novikov the conjecture, later proved by Shiota, that indeed the KP equation suffices to characterize the theta functions associated to Riemann surfaces.

As Mumford pointed out in several occasions, one may interpret the KP equation in terms of a geometrical property of Jacobian varieties. This property was discovered by A. Weil and can be described in the following way. On *any* Jacobian J there exists a point v such that:

$$\Theta \cap \{\Theta + v\} \subset \{\Theta + w\} \cup \{\Theta + w'\}$$

for some pair of points w, $w' \in J$ such that $w, w' \notin \{0, v\}$. It turns out that the KP equation can be viewed as the infinitesimal analogue at the above decomposition.

In the first four section of the present work we study this analogy. We start by illustrating Matsusaka's way of characterizing Jacobian varieties among all principally polarized abelian varieties. We then proceed to study Gunning's and Welters' criteria. These criteria are built on Matsusaka's criterium and are inspired by Fay's trisecant formula which in turn can be viewed as an analytical way of expressing Weil's decomposition. From Welters' criterium we pass to the KP hierarchy showing that the former is the geometrical translation of the latter.

In the last three sections we present the theory of the KP hierarchy based on the infinite dimensional Grassmannian and the τ function. Our main motivation in this context was to present a relatively short proof of the fact that the τ function satisfies the Hirota bilinear relation. This relation is one of the many incarnations of the KP hierarchy and it is the closest to its geometrical interpretation.

1. Riemann surfaces and Abelian varieties

Let us consider a compact Riemann surface C of genus g (we shall also call C a curve of genus g). Denote by $a_1, \ldots, a_g, b_1, \ldots, b_g$ a *symplectic basis* of the first homology group $H_1(C, \mathbf{Z})$. This means that the intersection matrix is of the form

$$\begin{pmatrix} a\cdot a & a\cdot b \\ b\cdot a & a\cdot a \end{pmatrix} = \begin{pmatrix} 0 & I \\ -I & 0 \end{pmatrix}$$

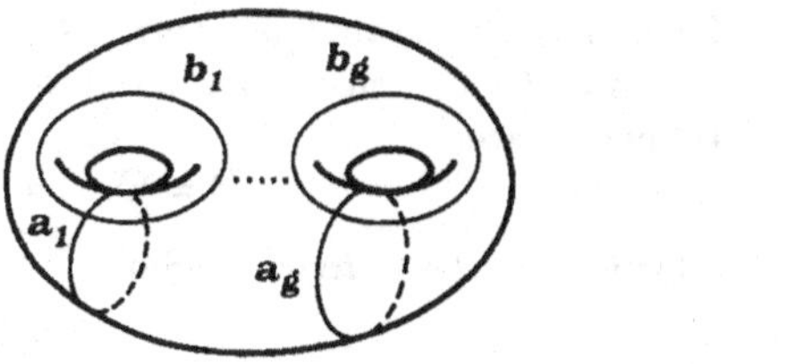

We shall denote by $\omega_1, \ldots, \omega_g$ a basis of the $\mathbf{C}$ vector space of holomorphic differentials n C. The complex $g \times 2g$ matrix

1.1) $$\Omega = \left(\int_{a_i} \omega_j \ , \ \int_{b_i} \omega_j \right)$$

called the *period matrix* of C. As it was observed by Riemann it is possible to choose the asis $\omega_1, \ldots \omega_g$ in such a way that the period matrix is of the form

1.2) $$\Omega = (I, \tau)$$

here τ is a *symmetric matrix with positive definite imaginary part*:

$$\tau = {}^t\tau, \ \mathrm{Im}\, \tau > 0.$$

n particular the column vectors of the period matrix generate a lattice Λ_τ in $\mathbf{C}^g$ of maximal nk. One can therefore attach the complex torus $\mathbf{C}^g/\Lambda_\tau$ to the Riemann surface C. This mplex torus is called the *Jacobian of C* and it is denoted by $J(C)$. The jacobian of C in fact much more than just a complex torus. It is a *principally polarized abelian variety* riefly p.p.a.v.). Let us pause for a while to talk about them.

An *abelian variety* is a complex torus which is also an algebraic variety. The datum of an mple line bundle L an abelian variety X is what is called a *polarization*. The polarization said to be *principal* if L admits only one non-zero holomorphic section up to a constant, e.: $\dim H^0(X, L) = 1$.

Let us consider the *generalized Siegel upper half plane* of dimension g, that is the set $\mathcal{H}_g$ $g \times g$ symmetric matrices with positive definite imaginary part. This is an open set in $\frac{g(g+1)}{2}$. Given a point $\tau \in \mathcal{H}_g$ consider the lattice $\Lambda_\tau \subset \mathbf{C}^g$ generated by the columns of e $g \times 2g$ matrix (I, τ), and form the complex torus $X_\tau = \mathbf{C}^g/\Lambda_\tau$.

Let $\zeta = (\zeta_1, \ldots, \zeta_g)$ and $\tau = (\tau_{ij})$ be coordinates in $\mathbf{C}^g$ and $\mathcal{H}_g$ respectively. The *iemann theta-function*

.3) $$\theta(\zeta, \tau) = \sum_{p \in \mathbf{Z}^g} \exp 2\pi i \left\{ \frac{1}{2}\, {}^t p \tau p + {}^t p \zeta \right\}$$

holomorphic in $\mathbf{C}^g \times \mathcal{H}_g$ and satisfies the quasi-periodicity property:

.4) $$\theta(\zeta + n + {}^t m\tau, \tau) = \exp 2\pi i \left\{ -\frac{1}{2}\, {}^t m \tau m - {}^t m \zeta \right\} \theta(\zeta, \tau)$$

for $n, m \in \mathbf{Z}^g$ (so that $n + {}^t m\tau \in \Lambda_\tau$). Fix τ, the quasi-periodicity of θ shows that the zeroes of θ are well defined modulo Λ_τ. Thus the *theta* – *function* gives rise to a divisor (hypersurface)

$$\Theta = \Theta(\tau) = \{\zeta \in X_\tau : \theta(\zeta, \tau) = 0\} \subset X_\tau.$$

This is the "*theta-divisor*". Since

(1.5) $$\theta(\zeta, \tau) = \theta(-\zeta, \tau)$$

the theta-divisor is *symmetric* with respect to the "-1" involution on X_τ:

(1.6) $$\Theta = -\Theta.$$

It is a fundamental result at Fourier analysis that:

(1.7) *Any p.p.a.v* (X, L) *is of the form* $(X, L) = (X_\tau, \Theta(\tau))$ *for some* $\tau \in \mathcal{H}_g$.

Two p.p.a.v. (X, L) and (X', L') are said to be *isomorphic* if there exists an isomorphism $\varphi : X \to X'$ s.t. $\varphi^* L' = L$. The preceeding result can be completed as follows.

(1.8) *Let* $\tau, \tau' \in \mathcal{H}_g$. *Then the p.p.a.v. corresponding to* τ and τ' *are isomorphic if and only if there exists* $\delta = \begin{pmatrix} \alpha & \beta \\ \gamma & \delta \end{pmatrix} \in \mathrm{Sp}(2g, \mathbf{Z})$ *such that*

$$\tau' = \frac{\alpha\tau + \beta}{\gamma\tau + \delta}.$$

The moduli space of p.p.a.v. is then given by the quotient

$$\mathcal{A}_g = \mathcal{H}_g / \mathrm{Sp}(2g, \mathbf{Z})$$
$$\|$$
$$\{\text{p.p.a.v.}\}/_{\text{iso}}.$$

It turns out that $\mathcal{A}_g$ is a quasi-projective variety of dimension $\frac{1}{2}g(g+1)$ (cf. [I]).

Let us now consider the moduli space $\mathcal{M}_g$ of curves of genus $g > 1$.

$$\mathcal{M}_g = \{\text{genus } g \text{ Riemann surfaces}\}/_{\text{iso}}$$

As is well known [MF] $\mathcal{M}_g$ is a quasi-projective variety of dimension $3g-3$. It is convenient for us to consider a topological cover $\widetilde{\mathcal{M}}_g$ of $\mathcal{M}_g$. This will be the moduli space of pairs consisting of a curve C together with a symplectic basis of $H_1(C, \mathbf{Z})$, modulo isomorphisms. The construction of the period matrix defines a holomorphic map

$$\widetilde{T} \; : \; \widetilde{\mathcal{M}}_g \longrightarrow \mathcal{H}_g$$

$$[C, a_1, \ldots, a_g, b_1, \ldots, b_g] \mapsto \tau = \left(\int_{b_j} \omega_j \right)$$

The analytic subvariety $\widetilde{T}(\widetilde{\mathcal{M}}_g)$ is called the *jacobian locus* and it is denoted by $\mathcal{J}_g$

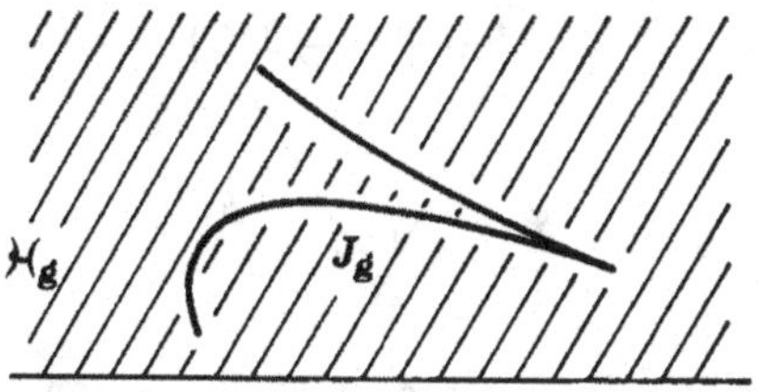

$$\left(\dim \mathcal{H}_g = \frac{1}{2}g(g+1),\ \dim\ \mathcal{J}_g = 3g-3\right)$$

We then have a commutative diagram

(1.9)
$$\begin{array}{ccc} \widetilde{\mathcal{M}}_g & \xrightarrow{\widetilde{T}} & \mathcal{H}_g \\ {\scriptstyle \pi}\downarrow & & \downarrow{\scriptstyle \pi'} \\ \mathcal{M}_g & \xrightarrow[T]{} & \mathcal{A}_g \end{array}$$

where π and π' are the natural projections and T is the map induced by $\widetilde{T}$. The map T is called the *Torelli map*, in view of the following theorem of Torelli

(1.10) **Theorem.** *The map T is injective.*

Torelli's theorem therefore says that there is a bijection

(1.11)
$$\mathcal{M}_g \leftrightarrow \mathcal{J}_g/\mathrm{Sp}(2g,\mathbf{Z})$$

For a more detailed description of this bijection cf. [OS]). A way of rephrasing Torelli's theorem is to say that a Riemann surface C can be completely reconstructed from its polarized jacobian

$(J(C),\Theta)$.

We shall now investigate the geometry of the theta divisor of a jacobian, we shall see how its geometry is linked to the curve it belongs to and how indeed it is possible from it to reconstruct the curve.

Consider first of all the *Abel–Jacobi map*

(1.12)
$$\begin{gathered} C \xrightarrow{u} J(C) = \mathbf{C}^g/\Lambda_\tau \\ p \longmapsto \int_{p_o}^{p} \vec{\omega} \end{gathered}$$

where $p_o \in C$ is a fixed point and $\vec{\omega} = (\omega_1,\ldots,\omega_g)$. The Abel–Jacobi map satisfies the following universal property (making $J(C)$ the *Albanese variety* of C):

(1.13) *For every abelian variety X and every morphism $\varphi : C \to X$ there exists a unique morphism $F : J(C) \to X$ such that $\varphi \doteq Fu$:*

$$\begin{array}{ccc} C & \xrightarrow{\quad u \quad} & J(C) \\ & {\scriptstyle\varphi}\searrow \quad \swarrow{\scriptstyle F} & \\ & X & \end{array}$$

If C_d denotes the d–fold symmetric product of C one can also define, by addition, the map

$$u_d = u : C_d \longrightarrow J(C) \tag{1.14}$$

$$P_1 + \cdots + P_d \longrightarrow u\,(P_1 + \cdots + P_d) = u(P_1) + \cdots + u(P_d)$$

Abel's theorem asserts that for $D \in C_d$

$$\boxed{u^{-1}(u(D)) = |D|} \tag{1.15}$$

where $|D|$ is the complete linear series determined by D. (This tells in particular that the Abel–Jacobi map $u : C \to J(C)$ gives an embedding of C into $J(C)$. We denote by W_d the image of C_d under u and we set

$$W_d^1 = \{x = u(D) \in W_d : \dim |D \geq 1\} \tag{1.16}$$

We are going to show the following decomposition. For every $p, q, r, s \in C$

$$\begin{aligned} W_{g-1} \cap \Big\{ W_{g-1} + u(p-q) \Big\} &= \Big\{ W_{g-2} + u(p) \Big\} \cup \Big\{ W_g^1 - u(q) \Big\} \\ &\subset \Big\{ W_{g-1} + u(p-r) \Big\} \cup \Big\{ W_{g-1} + u(s-q) \Big\} \end{aligned} \tag{1.17}$$

Let $D = P_1 + \cdots + P_{g-1} \in C_{g-1}$. Assume

i) $u(D) \in W_{g-1} \cap \{W_{g-1} + u(p-q)\}$

ii) $u(D) \notin W_g^1 - u(q)$.

By Abel's theorem i) implies that there exists $\Delta \in |D+q|$ s.t. $\Delta > p$. On the other hand ii) implies that $\dim |D+q| = 0$. Hence $\Delta = D + q$, so that $D > p$ and then

$$u(D) = u(D-p) + u(p) \in W_{g-2} + u(p).$$

This shows that

$$u(D) \in \{W_{g-2} + u(p)\} \cup \{W_g^1 - u(q)\}$$

The rest of (1.17) is just as easy. Set

$$\begin{aligned} X_1 &= W_{g-2} + u(p) \\ X_2 &= W_g^1 - u(q) \end{aligned} \tag{1.18}$$

From (1.17) we known that

$$W_{g-1} + u(p-r) \supset X_1$$

with some more effort one can show that, if C is *non hyperelliptic*, then

$$\{w : W_{g-1} + w \supset X_1\} = -u(C) + u(p) \tag{1.19}$$

As we shall see these remarks are among the basic ingredients in the proof at Torelli's theorem

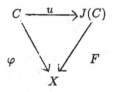

Another one is *Riemann's theorem* which asserts that there exists a point $\kappa \in J(C)$ such that

(1.20) $$\boxed{W_{g-1} = \Theta - \kappa}$$

and that, moreover

$$-2\kappa = u(K_C)$$

where K_C is a canonical divisor on C. (Incidentally Riemann's theorem tells us that the Jacobian of a curve is *indecomposable* in the sense that its theta divisor is irreducible).

From (1.17), (1.19), (1.20) we conclude that if C is a non–hyperelliptic curve with jacobian $J(C)$ and theta divisor Θ, then:

(1.21) *There exists a point* $v \in J(C)$ s.t.

$$\Theta \cap \{\Theta + v\} \subset \{\Theta + w\} \cup \{\Theta + w'\}$$

for some pair $\{w, w'\}$ with $w, w' \notin \{o, v\}$.

(1.22) *The intersection* $X = \Theta \cap \{\Theta + v\}$ *has two irreducible components* X_1 *and* X_2.

(1.23) **Torelli's theorem.** *The curve* $u(C)$ *is, up to* ± 1, *a translate of the locus:*

$$\{w \in J(C) : \Theta + w \supset X_1\}$$

This shows indeed that $u(C)$, and therefore C, can be reconstructed from Θ.

In the course of our study the decomposition (1.21) will appear time and again and in various disguises.

We end this section by recalling *Jacobi's theorem*, stating that the map

$$u : C_g \to J(C)$$

is *surjective*. The fact that the g–fold symmetric product of the curve is birationally equivalent to an abelian variety is a quite surprising one. Let us look at this more closely. Consider the theta divisor $\Theta \subset J(C)$. Given a point $e \in \Theta$, we known, by Riemann's theorem, that

$$e = u(D) + \kappa,$$

so that

(1.24) $$e = u(D + p_0) + k \qquad D + p_0 \in C_g.$$

Assume now that $e \notin \Theta$. Look at $\theta(u(p) - e)$ as a function on a fundamental polygon Π obtained by dissecting C along $2g$ simple curves meeting only at one point p_0 and generating $\pi_1(C, \mathbf{Z})$.

An elementary residue computation gives that $\theta(u(p) - e)$ has exactly g zeroes $p_1, \ldots, p_g$ and that

(1.25) $$e = u(p_1 + \cdots + p_g) + \kappa.$$

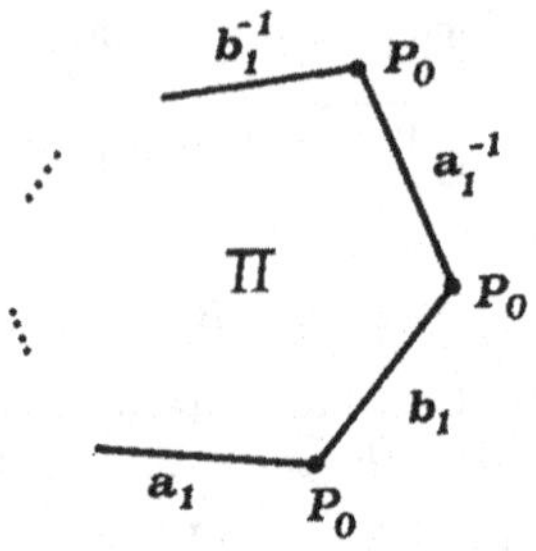

Observe that (1.24) and (1.25) give a constructive way to prove Jacobi's theorem. Namely, for $e \in J(C) \backslash \Theta$ one has

$$e = u\left(u^{*}\Theta_{e}\right) + \kappa \tag{1.26}$$

where Θ_e means $\Theta + e$.

Because of Abel's theorem the Abel–Jacobi map

$$u : C_d \rightarrow J(C)$$

drops to an *injection*

$$u : C_d/_\sim \rightarrow J(C)$$

where $\sim$ denotes linear equivalence. Jacobi's theorem tells us that *when $d \geq g$, the map u is actually an isomorphism.* In fact this is true in a strict sense, because, due to the universal property of the Abel–Jacobi map, and to the fact that $J(C)$ is indecomposable, one sees that *$C_d/_\sim$ is birationally equivalent to an abelian variety only if $d \geq g$.*

2. A criterion for Jacobian varieties

We shall talk about a geometrical way to recognize when a p.p.a.v. is the jacobian of a curve. The criterion we will explain is due to Matsusaka [M]. We shall state it in the form given by Hoyt [H] as to include the case of reducible curves.

All geometrical characterizations of Jacobians, among all abelian varieties, have more or less this form. One assumes that the given abelian variety X contains a curve (possibly singular reducible and non–reduced) having a certain numerical property. Then one shows that every component of C is smooth and that X is the jacobian of C, (if $C_1, \ldots, C_k$ are the components of C the jacobian of C is by definition $J(C_1) \times \cdots \times J(C_k)$).

The characterizing property that Matsusaka is using goes back to Poincaré. Let C be genus g curve. Let $\Gamma \subset J(C)$ be the image of C under the Abel–Jacobi map:

$$C \xrightarrow{u} u(C) = \Gamma \subset J(C).$$

Consider the $(g-1)$–fold self–intersection of Θ and let $[\Gamma]$ be the class of Γ. Then

$$\Theta^{(g-1)} = (g-1)!\,[\Gamma] \tag{2.1}$$

To prove this formula consider a basis

$$dx_I = dx_{i_1} \wedge dx_{i_2}$$

of $H^2(J(C),\mathbf{Z})$ where $x_1,\ldots,x_{2g}$ are real coordinates in $\mathbf{C}^g$ corresponding to the symplectic basis $a_1,\ldots,a_g,b_1,\ldots,b_g$ of $H_1(C,\mathbf{Z})=\Lambda$, and where we think of $J(C)$ as $\mathbf{C}^g/\Lambda$. One has to show that

$$\int_C u^* dx_I = \frac{1}{(g-1)!}\int_{J(C)} \Theta^{(g-1)} \wedge dx_I.$$

Set $\gamma_i = a_i, \gamma_{g+i} = b_i$. Since

$$\int_{\gamma_i} u^* dx_j = \delta_{ij}$$

u^*dX_j is Poincaré dual to γ_i. Therefore

$$\begin{aligned}\int_C u^*\left(dx_i \wedge dx_{g+i}\right) &= -\int_C u^*\left(dx_{g+i}\wedge dx_i\right)\\ &= (\gamma_i\cdot\gamma_{g+i}) = 1\end{aligned}$$

$$\int_C u^*\left(dx_i \wedge dx_{g+j}\right) = 0 \text{ if } |i-j| \neq 0.$$

On the other hand the fundamental class of Θ is given by

$$\xi = \sum dx_i \wedge dx_{g+i}$$

This is a purely topological fact that can be checked on an abelian variety which is the product of g elliptic curves). Then

$$\theta^{g-1} = (g-1)!\sum dx_1 \wedge dx_{g+1} \wedge\cdots\wedge \left(dx_i \widehat{\wedge dx}_{g+i}\right)\wedge\cdots\wedge dx_g \wedge dx_{2g}$$

so that

$$\int_{J(C)} \theta^{g-1} \wedge dx_I = \begin{cases} I = (i, g+i) \\ 0 \quad \text{otherwise}\end{cases}$$

proving (2.1). As a consequence of Poincaré formula and of the fact (cf. (1.26)) that Θ intersects Γ in g points one also gets

(2.2)
$$\boxed{\Theta^{(g)} = g!}$$

Consider now a general abelian variety X, of dimension n let $Z,Y \subset X$ be cycles of complementary dimension. Define an endomorphism of X by setting

(2.3)
$$\begin{aligned}\alpha(Z,Y):\ & X \longrightarrow X\\ & t \mapsto \alpha(Z,Y)(t) = S(Z\cdot(Y-Y_t))\end{aligned}$$

where S stands for the sum in X and Y_t stands for $Y+t$. By an elementary argument one shows that if Y is a *divisor* on X then

(2.4)
$$\alpha\left(Y^r, Y^{n-r}\right) = \frac{n-r}{n}\deg(Y^n) I$$

where I is the identity. The basic property of the endomorphism α is that if Y is a positive divisor and Δ a 1–cycle then

(2.5) $$\alpha(\Delta, Y) = 0 \Leftrightarrow \Delta \text{ is numerically equivalent to } 0.$$

Let us go back to the case of curves. Let $\Gamma \subset J(C)$ be the Abel Jacobi image at C. Then from (2.1), (2.2), (2.4) and (2.5) one gets

(2.6) $$\alpha(\Gamma, \Theta) = \alpha\left(\frac{\Theta^{g-1}}{(g-1)!}, \Theta\right) = \frac{1}{(g-1)!}\frac{1}{g}\deg\left(\Theta^{(g)}\right) I$$
$$= I$$

The criterion of Matsusaka–Hoyt is the following.

(2.7) Theorem *Let X be a p.p.a.v. of dimension $n > 1, Y \subset X$ a positive divisor, $\Gamma \subset X$ a positive 1-cycle. Then the following are equivalent*

1) *Γ is reduced, its reducible components are smooth, X is the jacobian of a curve C, the curve Γ is the image of C under the Abel-Jacobi map, and the divisor Y is the theta divisor on C.*
2) $\deg Y^{(n)} = n$, $[Y^{n-1}] = (n-1)![\Gamma]$
3) $\alpha(\Gamma, Y) = I$.

We shall not give the proof of this criterion but just a sketch of it. The fact that 2) implies 3) trivially follows from (2.4). The fact that 1) implies 3) is (2.6). The heart of the criterion is that 3) implies 1).

We assume for simplicity that Γ is irreducible. Let N be the normalization of Γ and

$$\varphi : N \dashrightarrow \Gamma \subset X$$

the normalization map. By the (Albanese) universal property of $J(N)$ one has a commutative diagram

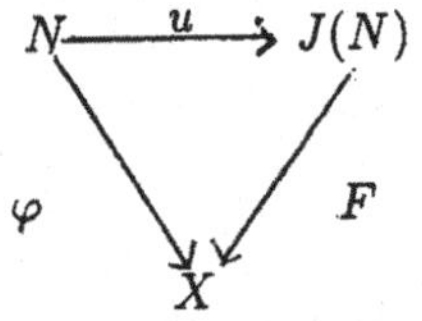

where u is the Abel–Jacobi map. The goal is to show that F is an isomorphism of p.p.a.v. We shall simply indicate how to see that F is an isomorphism of abelian varieties. Define the map

$$G : X \longrightarrow J(N)$$

by setting

$$G(t) = u\left(\varphi^*(\Gamma \cdot (Y - Y_t))\right)$$

(here $\Gamma \cdot (Y - Y_t)$ is a divisor in Γ, φ^* is the pull-back map, and $u_{\mathcal{D}} = u : \mathrm{Div}(N) \to J(N)$ is the Abel–Jacobi map on the group of divisors of N).

It is immediate to check that

$$F\,G = \alpha(\Gamma, Y)$$

Since, by hypothesis $\alpha(\Gamma, Y) = I$, it follows that G is injective. In order to show that G and hence F, is an isomorphism it suffices to show that $\dim J(N) = \dim X$. At this point we need the following fundamental property of abelian varieties.

2.8) **Theorem of the square.** *Let X be an abelian variety. For $a \in X$ denote by $T_a : X \to X$ the translation by $a : T_a(X) = X + a$. Let L be a line bundle on X. Let $a, b \in X$. Then*

$$T^*_{a+b}L \otimes L \cong T^*_a L \otimes T^*_b L \tag{2.9}$$

Equivalently if $Y \subset X$ is a divisor then

$$Y_{a+b} - Y_a - Y_b + Y \sim 0 \tag{2.10}$$

where $\sim$ denote the linear equivalence

An algebraic proof of this theorem is contained in Mumford book [M1]. Here we shall proceed analytically. First of all the *theorem of Appel-Humbert* (see [C]) says that *the isomorphism classes at line bundles on $X = \mathbf{C}^g/\Lambda$ are in one-to-one correspondence with pairs (H, ϱ) where H is integral on $\Lambda \times \Lambda$ and*

$$\varrho : \Lambda \to S^1$$

is a quasi-character, i.e.:

$$|\varrho(\lambda)| = 1$$

$$\varrho(\lambda)\varrho(\mu) = \varrho(\lambda + \mu)e^{\pi i E(\lambda,\mu)} = \pm\varrho(\lambda + \mu).$$

The line bundle $L(H, \varrho)$ corresponding to the pair (H, ϱ) in this correspondence is defined by

$$L(H, \varrho) = \mathbf{C}^g \times \mathbf{C}/\sim$$

where

$$(z, \xi) \sim \left(z + \lambda, \varrho(\lambda)e^{\pi H(z,\lambda) + \frac{\pi}{2}H(\lambda,\lambda)}\xi\right)$$

for $z \in \mathbf{C}^g$, $\xi \in \mathbf{C}$, $\lambda \in \Lambda$.

It is then obvious that the sections of $L(H, \varrho)$ correspond to holomorphic functions in $\mathbf{C}^g$ such that

$$\theta(z + \lambda) = \varrho(\lambda)e^{\pi H(z,\lambda) + \frac{\pi}{2}H(\lambda,\lambda)}\theta(z)$$

To prove the theorem of the square one then proceeded as follows. First of all given a line bundle $L = L(H, \varrho)$ and a point $a \in X$ one observes that the function

$$\widetilde{\theta}(z) = e^{-\pi H(z,a)}\theta(z + a)$$

corresponds to a section of $T^*_a L$, and satisfies the functional equation

$$\widetilde{\theta}(z + \lambda) = \varrho(\lambda)e^{2\pi i\, E(a,\lambda)}e^{\pi H(z,\lambda) + \frac{\pi}{2}H(\lambda,\lambda)}\widetilde{\theta}(z),$$

so that

$$T^*_a L(H, \varrho) = L\left(H, \varrho e^{2\pi i\, E(a,-)}\right)$$

One then observes that the theorem reduces to the identity:

$$\varrho(\lambda)e^{2\pi i\, E(a,\lambda)}\varrho(\lambda)e^{2\pi i\, E(b,\lambda)} = \varrho(\lambda)^2 e^{2\pi i\, E(a+b,\lambda)}$$

which is obvious.

Let us now go back to Matsusaka criterion.

We have

$$
\begin{array}{ccc}
 & \text{(Abel–Jacobi)} & \\
N & \xrightarrow{\ u\ } & J(N) \\
\text{(normalization map)}\ \varphi \searrow & G \ \nearrow\!\!\swarrow\ F & \text{(universal property of } J(N)) \\
 & \Gamma \subset X &
\end{array}
$$

$$F\,G = \alpha(\Gamma, Y) = I$$

From (2.4) we have

$$\alpha(Y^{n-1}, Y) = \frac{1}{n}\deg(Y^n)\cdot I, \quad n = \dim X$$

But $\alpha(\Gamma, Y) = I$ by hypothesis, so that, by the linearity of α

$$\alpha\left(Y^{n-1} - \frac{1}{n}\deg(Y^n)\cdot\Gamma\ ,\ Y\right) = 0.$$

But then by the fundamental property (2.5)

$$Y^{n-1} \equiv \frac{1}{n}\deg(Y^n)\cdot\Gamma$$

Intersecting with Y both sides of this equality we get

$$\deg\Gamma\cdot Y = n$$

In fact one can prove more. Namely that by translating Y on X and intersecting with one gets a *general* divisor of degree n on Γ. In other words one gets a rational map

$$
\begin{array}{ll}
X \cdots\cdots\triangleright N_n & \text{(n–fold symmetric product)} \\
a \longrightarrow \varphi^*(Y_a) &
\end{array}
$$

By the theorem of the square we get

$$\varphi^*\left(Y_{a+b} - Y_a - Y_b + Y\right) \sim 0$$

showing that $N_n/\sim$ is birationally equivalent to an abelian variety. But, as we pointed o at the end of the preceding section, this is possible only if

$$n \leq \dim J(N).$$

Q.E.

Matsusaka's criterion seems to have a weakness. In order to prove that an abelian varie X is the jacobian of a curve one needs to find a curve $\Gamma \subset X$. But we shall see that inde it is possible to make a quite efficient use of this criterion.

3. The trisecant formula and another criterion for Jacobian var eties

Let us go back to the decomposition of the theta divisor of a jacobian (cf. (1.17), (1.21) Let C be a curve of genus $g > 1$. Consider the Abel–Jacobi map

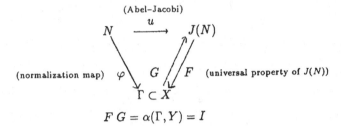

$$u : C \to J(C)$$

set $\Gamma = u(C)$. We know that for $p, q, r, s \in \Gamma$,

$$\boxed{\Theta \cap \Theta_{u(p-q)} \subset \Theta_{u(p-r)} \cup \Theta_{u(s-q)}}$$

etting

$$\alpha = u(s),\ \beta = u(q),\ \gamma = u(p),\ x = u(r) - \alpha - \beta - \gamma$$

ve can say that for $\alpha, \beta, \gamma \in \Gamma$ and for $x \in \Gamma - \alpha - \beta - \gamma$ we have

3.1)
$$\boxed{\Theta_\beta \cap \Theta_\gamma = \Theta_\alpha \cup \Theta_{-\alpha-x}}$$

Ve are now going to translate this in a more analytical way. For this we use a rather tandard result (cf. [W1]). Let X be a p.p.a.v. Assume X is indecomposable. Let $\widetilde{\Theta}$ be a lesingularization of its theta divisor then

3.2)
$$\dim H^\circ\left(\widetilde{\Theta}, \Theta_a\right) = 1$$

.e.

$$H^\circ\left(\widetilde{\Theta}, \Theta_a\right) = \theta_a|_\Theta \cdot \mathbf{C}$$

et α, β, γ and x be as above. Consider the exact sheaf sequences

3.3)
$$0 \to \mathcal{O}_X(\Theta_\beta) \xrightarrow{\theta_{-\beta-x}} \mathcal{O}_X(\Theta_\beta + \Theta_{-\beta-x}) \to \mathcal{O}_{\Theta_\beta}(\Theta + \Theta_{-\beta-x}) \to 0$$

3.4)
$$0 \to \mathcal{O}_{\Theta_\beta}(\Theta_{-\gamma-x}) \xrightarrow{\theta_\gamma} \mathcal{O}_{\Theta_\beta}(\Theta_{-\gamma-x} + \Theta_\gamma) \to \mathcal{O}_{\Theta_\beta \cap \Theta_\gamma}(\Theta_{-\gamma-x} + \Theta_\gamma) \to 0$$

y the theorem of the square we view $\theta_\alpha\theta_{-\alpha-x}$ as a section of $\mathcal{O}(\Theta_{-\gamma-x} + \Theta_\gamma)$. The relation 3.1) tells us in particular that this section vanishes on $\Theta_\beta \cap \Theta_\gamma$, so that by the cohomology equence of (3.4) and by (3.2) we get

.5)
$$\theta_\alpha\theta_{-\alpha-x}\Big|_{\Theta_\beta} = d\theta_\gamma\theta_{-\gamma-x}\Big|_{\Theta_\beta}, \quad d \in \mathbf{C}$$

gain by the theorem of the square we think of $\theta_\alpha\theta_{\gamma-x} - d\theta_\gamma\theta_{-\gamma-x}$ as a section of $(\Theta + \Theta_{-\beta-x})$, which by (3.5) vanishes on Θ_β. Therefore there exists a constant c such at

.6)
$$\theta_\alpha\theta_{-\alpha-x} = c\,\theta_\beta\theta_{-\beta-x} + d\theta_\gamma\theta_{-\gamma-x}$$

hich we also write as

$$\theta(z-\alpha)\theta(z+\alpha+x) = c\,\theta(z-\beta)\theta(z+\beta+x) + d\theta(z-\gamma)\theta(z+\gamma+x)$$

will be convenient to set $x = 2\zeta$ so that we get

.7)
$$\boxed{\begin{aligned}\theta(z-\alpha)\theta(z+2\zeta+\alpha) &= c\,\theta(z-\beta)\theta(z+2\zeta+\beta)\\ &\quad + d\theta(z-\gamma)\theta(z+2\zeta+\gamma)\end{aligned}}$$

This is the analytical form of the decomposition (3.1), it is called the *trisecant formula of Fay*. To understand the reason of this terminology we have to talk some about the Kummer variety and its geometry.

Let X be a p.p.a.v. and Θ its theta divisor. We are going to study the geometry of the linear system $|2\Theta|$. This is done via the *second-order theta function*. Given any half integer vector $n \in \frac{1}{2}\mathbf{Z}^g/\mathbf{Z}^g$, are defines *the second-order theta function with characteristics* $(n,0)$ as follows:

$$\theta[n](z,\tau) = \sum_{p\in\mathbf{Z}^g} \exp 2\pi i \left\{\frac{1}{2}\,{}^t(p+n)\tau(p+n) + {}^t(p+n)z\right\}. \tag{3.8}$$

Using the Riemann–Roch theorem and a direct computation, one can see that the 2^g functions

$$\widehat{\theta}[n](z,\tau) = \theta[n](2z,2\tau), \quad n \in \frac{1}{2}\mathbf{Z}^g/\mathbf{Z}^g$$

form a basis for the vector space of sections of $\mathcal{O}(\Theta)$. The morphism defined by the linear system $|2\Theta|$ is then given by

$$\begin{aligned} \vec{\theta} : X &\longrightarrow \mathbf{P}^N,\ N = 2^g - 1 \\ \zeta &\longrightarrow \vec{\theta}(\zeta) = [\ldots, \widehat{\theta}[n](\zeta,\tau), \ldots]. \end{aligned} \tag{3.9}$$

This is a two–to–one morphism. Its image is the so called *Kummer variety* $K(X)$. The Kummer variety has degree $2^g g!$ and it is smooth except at the images at the points a order two of X, where it is singular. Many identities involving the theta function can be geometrically interpreted in terms of the Kummer variety by means of a fundamental identit discovered by Riemann:

$$\theta(z+\zeta)\theta(z-\zeta) = \sum_{n\in\frac{1}{2}\mathbf{Z}^g/\mathbf{Z}^g} \widehat{\theta}[n](z)\widehat{\theta}[n](\zeta) \tag{3.10}$$

Using *Riemann's identity* (3.10) the trisecant formula (3.7) becomes

$$\sum_{n\in\frac{1}{2}\mathbf{Z}^g/\mathbf{Z}^g} \left[\widehat{\theta}[n](\zeta+\alpha) - c\widehat{\theta}[n](\zeta+\beta) - d\widehat{\theta}[n](\zeta+\gamma)\right] \cdot \widehat{\theta}[n](z+\zeta) = 0$$

Therefore by the linear independence of the $\widehat{\theta}[n]$'s the trisecant formula is equivalent to th system

$$\begin{gathered} \widehat{\theta}[n](\xi+\alpha) - c\widehat{\theta}[n](\zeta+\beta) - d\widehat{\theta}[n](\zeta+\gamma) = 0 \\ \zeta \in \frac{1}{2}(\Gamma - \alpha - \beta - \gamma), \quad n \in \frac{1}{2}\mathbf{Z}^g/\mathbf{Z}^g \end{gathered} \tag{3.11}$$

where $\frac{1}{2}$ stands for the inverse image of the multiplication by 2 map on X. We can writ the system (3.11) in vector notation:

$$\vec{\theta}(\zeta+\alpha) = c\vec{\theta}(\zeta+\beta) + d\vec{\theta}(\zeta+\gamma), \quad \zeta \in \frac{1}{2}(\Gamma - \alpha - \beta - \gamma) \tag{3.12}$$

r equivalently as

(3.13) $$\boxed{\vec\theta(\xi+\alpha)\wedge\vec\theta(\xi+\beta)\wedge\vec\theta(\xi+\gamma)=0, \quad \zeta\in\frac{1}{2}(\Gamma-\alpha-\beta-\gamma)}$$

This has the following geometrical interpretation. For any choice of points $\alpha,\beta,\gamma\in\Gamma$, the urve $\frac{1}{2}(\Gamma-\alpha-\beta-\gamma)$ parametrires trisecant lines to the Kummer variety of $J(C)$. In fact 3.13) says that for any $\zeta\in\frac{1}{2}(\Gamma-\alpha-\beta-\gamma)$ the three points $\vec\theta(\zeta+\alpha)$, $\vec\theta(\zeta+\beta)$, $\vec\theta(\zeta+\gamma)$ re collinear:

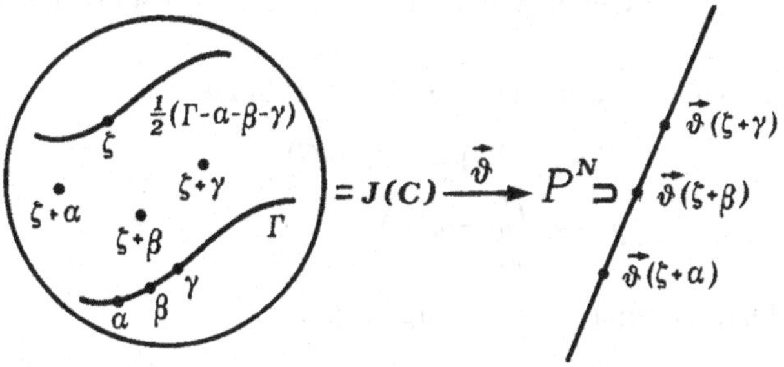

One should keep in mind that, as we just showed, the decomposition (3.1) of the theta ivisor, the analytical formula (3.7) for Riemann's theta divisor and the geometrical interpreation (3.13) in terms of trisecants to the Kummer variety, are different aspect of the same henomenon.

Gunning in [G] showed that the trisecant formula can be used to characterize jacobians mong all abelian varieties. He shows that *an indecomposable principally polarized abelian ariety (i.p.p.a.v.) X is a jacobian if and only if there exists an irreducible curve Γ in X uch that for general points α,β,γ on Γ and for every point ζ on $\frac{1}{2}(\Gamma-\alpha-\beta-\gamma)$ the three oints $\vec\theta(\zeta+\alpha)$, $\vec\theta(\zeta+\beta)$, $\vec\theta(\zeta+\gamma)$ are collinear.*

More precisely given three points α,β,γ on X, Gunning introduces the subvariety $_{\alpha,\beta,\gamma}\subset X$ defined by

.14) $$\widetilde V_{\alpha,\beta,\gamma}=\Big\{\zeta\in X:\vec\theta(\zeta+\alpha)\wedge\vec\theta(\zeta+\beta)\wedge\vec\theta(\zeta+\gamma)=0\Big\},$$

en, using Matsusaka's criterion, he shows that *X is the jacobian of a curve if and only if e following two condition hold*

.15) $$\dim_{-\alpha-\beta}2\widetilde V_{\alpha,\beta,\gamma}>0 \quad for \quad some \quad \alpha,\beta,\gamma \quad on \quad X$$

.16) *There is no complex multiplication on X mapping $\beta-\alpha$ and $\gamma-\alpha$ into 0.*

Ioreover if these conditions are satisfied then $\Gamma=2V_{\alpha,\beta,\gamma}$ turns out to be a smooth curve nd X is the Jacobian of Γ.

The remarkable aspect of this result is that the scheme $\widetilde V_{\alpha,\beta,\gamma}$ *is completely defined in rms of theta function and therefore lives* on any *abelian variety X*. In general $\widetilde V_{\alpha,\beta,\gamma}$ is zero–

dimensional and is one-dimensional exactly when X is a jacobian. This is an improvement on Matsusaka's criterion in that it provides a concrete candidate for the curve Γ, a candidate one can try to compute with!

We shall give a proof of a modified version of Gunning's criterion (one in which, by the way, the unpleasant condition (3.16) disappears). The modification we have in mind consists in letting the three points α, β and γ tend to zero. The resulting criterion was proved by Welters [W2]. The idea of using this limiting procedure is due to Mumford, and it leads directly to the Kodomcev-Petvieshvili equation (K.P.). It is in fact Mumford who pointed out, in several occasion, the relation between the K.P. equation and the various features of the decomposition (3.1) (cf. [MF], [M3], [M4]).

To state Welter's version of Gunning's criterion we start with some heuristic considerations.

Let C be a curve of genus g. Consider the Abel–Jacobi map:

$$u : C \longrightarrow u(C) = \Gamma \subset J(C) = X$$
$$p \longrightarrow \int_{p_0}^{p} \vec{\omega} = u(P)$$

Let ϵ be a local coordinate around p_0 vanishing on p_0.

Set

$$\int_{p_0}^{p} \vec{\omega} = 2\left(\vec{W}^1 \epsilon + \vec{W}^2 \epsilon^2 + \dots\right), \tag{3.17}$$

so that a parametric expression of the curve $\frac{1}{2}\Gamma$ near 0 will be

$$\vec{\zeta(\epsilon)} = \epsilon \vec{W}^1 + \epsilon^2 \vec{W}^2 + \dots \tag{3.18}$$

(explaining the factor of 2 in (3.17)). We now introduce some notation. Let

$$D = (D_1, D_2, \dots)$$

be a collection of *constant* vector fields in $\mathbf{C}^g$. Write

$$D(\epsilon) = \sum_{i \geq 1} D_i \epsilon^i$$

Define differential operators $\Delta_j = \Delta_j(D)$ by

$$e^{D(\epsilon)} = \sum_{j \geq 0} \Delta_j(D) \epsilon^j,$$

so that:

$$\Delta_0 = I, \quad \Delta_1 = D_1, \quad \Delta_2 = \frac{1}{2} D_1^2 + D_2, \quad \Delta_3 = \frac{1}{3!} D_1^3 + D_1 D_2 + D_3, \dots$$

$$\Delta_s = \sum_{i_1 + 2i_2 + \dots + s i_s = s} \frac{1}{i_1! \dots i_s!} D_1^{i_1} \dots D_s^{i_s}.$$

Now given a germ of a holomorphic curve near the origin in $\mathbf{C}^g$

$$\vec{\zeta(\epsilon)} = \epsilon \vec{W}^1 + \epsilon^2 \vec{W}^2 + \dots, \qquad \vec{W}^i \equiv (W_1^i, \dots, W_g^i)$$

ve set

$$D_i = \sum_{j=1}^{g} W_j^i \frac{\partial}{\partial \zeta_j}$$

vhere $\zeta = (\zeta_1, \ldots \zeta_g)$ are coordinates in $\mathbf{C}^g$. For a holomorphic function $f(\zeta)$ in $\mathbf{C}^g$ we hen have

3.19)
$$\boxed{f(\zeta + \zeta(\vec{\epsilon})) = e^{D(\epsilon)} f(\zeta) = \sum_{j \geq 0} \Delta_j(f)\epsilon^j}$$

et us go back to the trisecant formula

3.20)
$$\vec{\theta}(\zeta + \alpha) \wedge \vec{\theta}(\zeta + \beta) \wedge \vec{\theta}(\zeta + \gamma) = 0$$

here $\alpha, \beta, \gamma \in \Gamma$. We can write

$$\alpha = 2\zeta(\vec{\epsilon}),$$

nd since

$$\begin{aligned} \Delta_1(2D) &= 2D_1 = 2\Delta_1 \\ \Delta_2(2D) &= 2D_1^2 + 2D_2 \\ &= \text{(by definition)} \quad 2\overline{\Delta}_2, \end{aligned}$$

e have

$$\vec{\theta}(\zeta + 2\zeta(\vec{\epsilon})) = \vec{\theta}(\zeta) + 2\epsilon\Delta_1\vec{\theta}(\zeta) + 2\epsilon^2\overline{\Delta}_2\vec{\theta}(\zeta) + \ldots$$

herefore when α, β, γ tend to zero along Γ the trisecant formula (3.20) becomes

.21)
$$\boxed{\vec{\theta}(\zeta) \wedge \Delta_1\vec{\theta}(\zeta) \wedge \overline{\Delta}_2\vec{\theta}(\zeta) = 0, \quad \zeta \in \frac{1}{2}\Gamma}$$

he geometrical interpretation of this formula is clear. For every point $\zeta \in \frac{1}{2}\Gamma$, there exists line $l \subset \mathbf{P}^N$ which is an *inflectionary trisecant* to the Kummer variety. More precisely the ne l is such that

$$(\vec{\theta})^*(l) = \zeta + Y$$

here $Y \subset X$ is the subscheme given by

$$\begin{aligned} \text{Spec } \mathbf{C}[\epsilon]/\epsilon^3 &\hookrightarrow X \\ \epsilon &\longmapsto \epsilon\vec{W}^1 + \epsilon^2\vec{W}^2. \end{aligned}$$

hus when X is a Jacobian the image under the Kummer map $\vec{\theta}$ of the curve $\frac{1}{2}\Gamma \subset X$ is a cus of inflectionary points for the Kummer variety

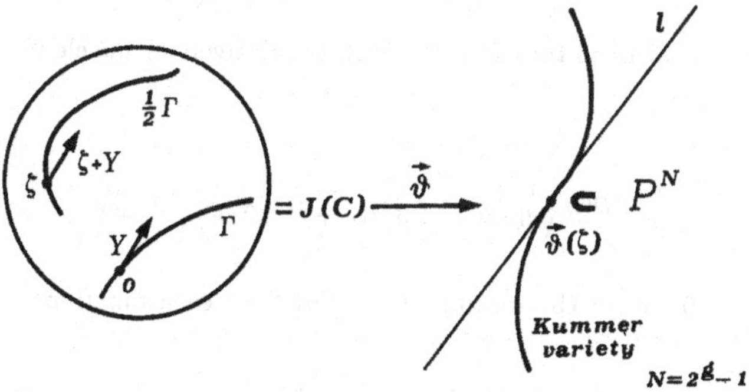

This, if one thinks about it, is a very unlikely situation. One has a g-dimensional variety namely the Kummer variety $K(X)$, sitting in a huge projective space, namely $\mathbf{P}^N$, $N = 2^g$ and this variety possesses a large number of trisecant and inflectionary lines. As we shall see this peculiarity characterizes jacobians among all abelian varieties.

Let now X be *any* p.p.a.v., let D_1 and D_2 be constant vector fields on X. We define the subscheme

$$\widetilde{V}_{D_1,D_2} = \left\{\zeta \in X : \vec{\theta}(\zeta) \wedge \Delta_1\vec{\theta}(\zeta) \wedge \overline{\Delta}_2\vec{\theta}(\zeta) = 0\right\}$$

We can now state Welter's version of Gunning's criterion.

(3.22) *Let X be an indecomposable* p.p.a.v. *of dimension g. Then X is a Jacobian if and only if there exist constant vector fields $D_1 \neq 0$, D_2 such that*

$$\dim_0 \widetilde{V}_{D_1,D_2} > 0. \tag{3.23}$$

Moreover if this condition holds, then $\widetilde{V}_{D_1,D_2}$ is a smooth curve and X is the jacobian of $\Gamma = 2\widetilde{V}_{D_1,D_2}$

We would like to present a sketch of a proof of Welter's criterion. Before giving this sketch we will analyse in some detail the meaning of the condition (3.23). We shall first translate it in more analytical terms and then reinterpret back our results in a geometrical way.

The first remark is that the subscheme $\widetilde{V}_{D_1,D_2}$ *coincides up to second order with the germ*

$$\epsilon \to \epsilon W^1 + \epsilon^2 W^2 \tag{3.24}$$

where $D_i = \sum_{j=1}^g W_j^i \frac{\partial}{\partial \zeta_i}$ $i = 1, 2$. So that, in particular

$$\vec{\theta}(\epsilon W^1 + \epsilon^2 W^2) \wedge \Delta_1\vec{\theta}(\epsilon W^1 + \epsilon^2 W^2) \wedge \overline{\Delta}_2\vec{\theta}(\epsilon W^1 + \epsilon^2 W^2) = 0 \bmod \epsilon^3.$$

Now the condition that $\widetilde{V}_{D_1,D_2}$ should be at least 1-dimensional at 0 means that there must be an analytical germ of a curve at zero contained in $\widetilde{V}_{D_1,D_2}$. Since, as we just remarked this germ should coincide with (3.24) up to second order, we shall write it in the form

$$\zeta(\vec{\epsilon}) = \epsilon W^1 + \epsilon^2 W^2 + \epsilon^3 W^3 + \dots$$

The condition (3.23) can now be written in the form

$$\vec{\theta}(\zeta(\vec{\epsilon})) \wedge \Delta_1\vec{\theta}(\zeta(\vec{\epsilon})) \wedge \overline{\Delta}_2\vec{\theta}(\zeta(\vec{\epsilon})) \equiv 0 \tag{3.25}$$

This can be written in the form

$$e^{D(\epsilon)} \left(\vec{\theta}(\zeta) \wedge D_1\vec{\theta}(\zeta) \wedge \overline{\Delta}_2\vec{\theta}(\zeta)\right)\Big|_{\zeta=0} = 0,$$

which in turn is equivalent to the existence of three relatively prime elements $d(\epsilon), c(\epsilon), b(\epsilon)$ $\mathbf{C}\{\epsilon\}$.

Such that

$$e^{D(\epsilon)} \left(d(\epsilon)\vec{\theta} + c(\epsilon)D_1\vec{\theta} + b(\epsilon)\overline{\Delta}_2\vec{\theta}\right)\Big|_{\zeta=0} = 0. \tag{3.26}$$

Computing at $\epsilon = 0$ we get (by recalling that $\vec{\theta}[n]$ is an even function)

$$\left(d(0)\vec{\theta}(0)+b(0)\overline{\Delta}_2\vec{\theta}(0)\right)\Big|_{\zeta=0}=0,$$

r equivalently (again by evenness)

(3.27) $$d(0)\vec{\theta}(0)+b(0)D_1^2\vec{\theta}(0)=0$$

ow a theorem of Wirtinger [I] asserts that when the theta divisor is *indecomposable* then the ectors $\vec{\theta}(0)$ and $\frac{\partial^2}{\partial\zeta_i\partial\zeta_j}\vec{\theta}(0)$ $i,j=1,\dots,g$ are linearly independent. From this and (3.27) follows that $d(0)=b(0)=0$. In particular since d,b,c are relative prime, this means that is invertible so that we may assume $c=1$. We now compute the coefficient of ϵ on the left de of (3.26). This is $\Delta_1^2\vec{\theta}(0)+b_1\Delta_1^2\vec{\theta}(0)$, where $b(\epsilon)=\sum_{i\geq 1}b_i\epsilon^i$. Thus $b_1=-1$. Changing ariables we can assume $b(\epsilon)=-1$. This amounts to changing each D_i by adding to it a near combination of $D_1,\dots,D_{i-1}$ which we may do. Now set

$$d(\epsilon)=\sum_{i=1}^{\infty}d_{i+1}\epsilon^i$$

he equation (3.26) mod ϵ^3 is $\epsilon\vec{\theta}(0)(d_2+\epsilon d_3)=0$ hence $d_2=d_3=0$. Therefore the equation .26), expressing condition (3.23), becomes

(3.28) $$\boxed{e^{D(\epsilon)}\left[\Delta_1-\epsilon\overline{\Delta}_2+d(\epsilon)\right]\vec{\theta}\Big|_{\zeta=0}=0}$$

et us now recall how we interpreted the decomposition (3.1) reaching the trisecant formula .7) and then, via Riemann's identity, the formula (3.13):

(3.1) $\Theta_\beta\cap\Theta_\gamma=\Theta_\alpha\cup\Theta_{-\alpha-x}$ $\qquad\alpha,\beta,\gamma\in\Gamma$

(via exact sheaf sequence)

(3.7) $\theta(z-\alpha)\theta(z+2\zeta+\alpha)$ "trisecant formula"

$=c\theta(z-\beta)\theta(z+2\zeta+\beta)+d\theta(\zeta-\gamma)\theta(z+2\zeta+\gamma)\qquad\zeta\in\frac{1}{2}(\Gamma-\alpha-\beta-\gamma)$

$\left(\text{via Riemann identity: } \theta(x-y)\theta(x+y)=\sum\hat{\theta}[n](x)\hat{\theta}[n](y)\right)$

(3.13) $\vec{\theta}(\zeta+\alpha)\wedge\vec{\theta}(\zeta+\beta)\wedge\vec{\theta}(\zeta+\gamma)=0\qquad\boxed{\zeta=\zeta(\epsilon)\in\frac{1}{2}\Gamma}$

ow we start from the infinitesimal analog of (3.13), namely

$$\vec{\theta}(\zeta(\epsilon))\wedge D_1\vec{\theta}(\zeta(\epsilon))\wedge\overline{\Delta}_2\vec{\theta}(\zeta(\epsilon))=0$$

d we go backwards. Actually we may start from (3.28), and using Riemann's identity we t

(3.29) $$e^{D(\epsilon)}\left[D_1-\epsilon\overline{\Delta}_2+d(\epsilon)\right]\theta(z+\zeta)\theta(z-\zeta)\Big|_{\zeta=0}=0$$

is is the equivalent of the trisecant formula (3.7). We are now going to deduce *two* analogues (3.1).

First analogue of the decomposition (3.1).

Recalling the definition of $D(\epsilon)$ and $e^{D(\epsilon)}$ and expanding in powers of ϵ we get

$$e^{D(\epsilon)}\left[D_1 - \epsilon\overline{\Delta}_2 + d(\epsilon)\right] = \sum_{s\geq 0}\left(\Delta_s\Delta_1 - \Delta_{s-1}\overline{\Delta}_2 + \sum_{i=3}^{s} d_{i+1}\Delta_{s-i}\right)\epsilon^s \tag{3.30}$$

Therefore (3.29) is equivalent to the system of differential equations

$$\boxed{\left[\Delta_s\Delta_1 - \Delta_{s-1}\overline{\Delta}_2 + \sum_{s=3}^{s} d_{i+1}\Delta_{s-i}\right]\theta(z+\zeta)\theta(z-\zeta)\Big|_{\zeta=0} = 0.} \tag{3.31}$$

The first non-trivial equation among these correspond to the value $s = 3$:

$$\left[\Delta_3\Delta_1 - \Delta_2\overline{\Delta}_2 + d\right]\theta(z+\zeta)\theta(z-\zeta)\Big|_{\zeta=0} = 0, \quad d = d_4$$

or more explicitly

$$\left[D_1^4 + 3D_1D_3 - D_2^2 + d\right]\theta(z+\zeta)\theta(z-\zeta)\Big|_{\zeta=0} = 0 \tag{3.32}$$

This is the so called *Kadomtsev–Petviashvili* equation (K.P) [Actually the classical form the K.P. equation is

$$\frac{\partial}{\partial x}\left(2\,u_{xxx} + 3u\,u_x - u_t\right) + 3u_{yy} = 0 \tag{3.33}$$

but this reduces to (3.32) under the substitution

$$u(x,y,t;z) = \frac{\partial^2}{\partial x^2}\log\theta\left(x\vec{W}^1 + y\vec{W}^2 + t\vec{W}^3 + Z\right)]$$

Let us write down (3.32) in an explicit form. We get

$$\begin{aligned} &D_1^4\theta(z)\cdot\theta(z) - 4D_1^3\theta(z)\cdot D_1\theta(z) + 3\left(D_1^2\theta(z)\right)^2 - 3\left(D_2\theta(z)\right)^2 \\ &+3D_2^2\theta(z)\cdot\theta(z) + 3D_1\theta(z)\cdot D_3\theta(z) - 3D_1D_3\theta(z)\cdot\theta(z) + d\theta(z)\cdot\theta(z) = 0 \end{aligned} \tag{3.34}$$

Consider now the theta–divisor $\Theta \subset X$. The function $D_1\theta$, when restricted to Θ, can considered as a section of the line bundle $\mathcal{O}_\Theta(\Theta)$. Its zero locus defines a divisor on Θ whi we denote by $D_1\Theta$, so that

$$D_1\Theta = \{\zeta \in X : \theta(\zeta) = D_1\theta(\zeta) = 0\}$$

Now let us look at the K.P. equation (3.34) and let us restrict it to $D_1\Theta$, we get

$$\left[\left(D_1^2\theta\right)^2 - \left(D_2\theta\right)^2\right]\Big|_{D_1\Theta} = 0$$

or equivalently

$$\left[\left(D_1^2+D_2\right)\theta\right]\left[\left(D_1^2-D_2\right)\theta\right]\Big|_{D_1\Theta}=0$$

We then see that if the K.P. equation is satisfied we have

(3.35) $$\boxed{\{\theta=0\}\cap\{D_1\theta=0\}\subset\left\{\left(D_1^2+D_2\right)\theta=0\right\}\cup\left\{\left(D_1^2-D_2\right)\theta=0\right\}}$$

which is an infinitesimal version of the decomposition (3.1)

Second analogue of the decomposition (3.1)
Let us go back to formula (3.29) and let us write it in the form

$$\left[D_1-\epsilon\overline{\Delta}_2+d(\epsilon)\right]\theta(z+\zeta(\epsilon)+\zeta)\theta(z-\zeta(\epsilon)-\zeta)\Big|_{\zeta=0}=0$$

Since the point z is arbitrary we may as well substitute $z+\zeta(\epsilon)$ to z and obtain

$$\left[D_1-\epsilon\overline{\Delta}_2+d(\epsilon)\right]\theta(z+2\zeta(\epsilon)+\zeta)\theta(z-\zeta)\Big|_{\zeta=0}=0$$

Expanding and setting $y=2\zeta(\epsilon)$ we get

$$\begin{aligned}&D_1\theta(z+y)\cdot\theta(z)-\theta(z+y)D_1\theta(z)-\epsilon D_1^2\theta(z+y)\theta(z)\\&+\epsilon D_1\theta(z+y)\cdot D_1\theta(z)-\epsilon\theta(z+y)\cdot D_1^2\theta(z)-\epsilon D_2\theta(z+y)\cdot\theta(z)\\&+\epsilon\theta(z+y)\cdot D_2\theta(z)+d(\epsilon)\theta(z+y)\theta(z).\end{aligned}$$

Setting $\theta(z+y)=\theta(z)=0$ we get our second analogue of (3.1) namely

(3.36) $$\boxed{\Theta\cap\Theta_{-y}\subset D_1\Theta\cup(D_1\Theta)_{-y}}\qquad y=2\vec{\zeta}(\epsilon)$$

and by symmetry

(3.37) $$\boxed{\Theta\cap\Theta_{y}\subset D_1\Theta\cup(D_1\Theta)_{y}}\qquad y=2\vec{\zeta}(\epsilon)$$

We are now going to sketch a proof of Welter's criterion (3.22). Suppose then that $\lim_0 \widetilde{V}_{D_1,D_2}>0$. Let Γ be the irreducible component of $2\widetilde{V}_{D_1,D_2}$ passing through the origin.

From (3.36) it follows that for every $y\in\Gamma$

$$\Theta\cap\Theta_y=X^1+X_y^2$$

where X^1,X^2 are cycles of codimension 2. The X_y^2 cover Θ as y moves on Γ so that $[X^2]*[\Gamma]=$ const $[\Theta]$, where $*$ is the Pontryagin product. Similarly

$$\Theta\cap\Theta_{-y}=X_{-y}^1+X^2$$

implies that $[X^1]*[\Gamma]=[X^1]*[\Gamma]=$ const $[\Theta]$. It follows that $[\Theta^2]*[\Gamma]=$ (const) $[\Theta]$. But

$$[\Theta^2]* : H^{2g-2}(X) \to H^2(X)$$

is an isomorphism. This gives that

$$[\Gamma] = (\text{const})[\Theta^{g-1}]/(g-1)!$$

It follows that (cf. 2.4))

$$\boxed{\alpha(\Gamma, \Theta) = (\text{const})I} \tag{3.38}$$

If we could prove that this constant is 1 we would just apply *Matsusaka's criterion* (2.7) an we would be done.

Let $\varphi : N \to \Gamma \subset X$ be the normalization map. Set

$$X = \mathbf{C}^n/\Lambda_\tau, \quad \tau \in \mathcal{H}_n, \quad J(N) = \mathbf{C}^g/\Lambda_\sigma, \quad \sigma \in \mathcal{H}_g$$

We then have by universality

$$\begin{array}{ccc} N & \xrightarrow{\quad u \quad} & J(N) \\ & {\scriptstyle \psi} \searrow \qquad \swarrow {\scriptstyle F} & \\ & X & \end{array}$$

Let C be the matrix of F, so that $C : \mathbf{C}^g \to \mathbf{C}^n$. Since $C\Lambda_\sigma \subset \Lambda_\tau$ there must be an integr $2n \times 2g$ matrix

$$Q = \begin{pmatrix} K & L \\ M & N \end{pmatrix}$$

such that

$$C(I, \sigma) = (I, \tau)Q$$

So that

$$\left\{ \begin{aligned} C &= K + \tau M \\ C\sigma &= L + \tau N \end{aligned} \right. \tag{3.39}$$

Consider now Matsusaka's map:

$$\alpha(\Gamma, \Theta) : \; X \xrightarrow{\chi} Pic^0(N) \xrightarrow{\psi^*} Pic^0(N) \xrightarrow{u} J(N) \xrightarrow{F} X$$

$$e \longmapsto \Big(\Theta_e - \Theta\Big) \longmapsto \psi^*\Big(\Theta_e - \Theta\Big) \longmapsto u\Big(D_e - D\Big) \longmapsto Fu\Big(D_e - D\Big)$$

$$\psi^*\Big(\Theta_e - \Theta\Big) = D_e - D$$

Since we know that $\alpha(\Gamma, \Theta)$ is a multiple of the identity it suffices to show that

$$\alpha(\Gamma, \Theta)(2W^1) = 2W^1$$

where $2W^1$ is the tangent vector to Γ at $0 \in X$. This is a highly non-trivial computatio but a very explicit one. To give an example one can compute $\alpha(\Gamma, \Theta)$ as a matrix:

$$\alpha(\Gamma, \Theta) = C({}^tN - \sigma\, {}^tM)$$

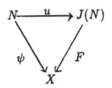

and this is achieved by observing that

$$\alpha(\Gamma,\Theta)(e)=\frac{1}{2\pi i}\int_{\partial\Pi}\vec{\varphi}(z)d\log\theta(\vec{\varphi}(z)-e)dz$$

when Π is a fundamental polygon for N. One then uses in a very explicit way the equations (3.25) to show that

$$C({}^tN-\sigma\,{}^tM)W^1=W^1.$$

. Characterizing Jacobians via the K.P. equation

Let us summarize what we did in the preceding section. We proved Welter's version of Gunning's criterion and we translate it into an analytical form. Namely we showed

(4.1) *An indecomposable* p.p.a.v X *is a jacobian if and only if there exist constant vector fields* $D_1\neq 0, D_2, D_3,\ldots$ *and constants* $d_4, d_5, d_6,\ldots$ s.t.

(4.2)
$$\left[\Delta_s\Delta_1-\Delta_{s-1}\overline{\Delta}_2+\sum_{i=3}^{s}d_{i+1}\Delta_{s-i}\right]\theta(z+\zeta)\theta(z-\zeta)|_{\zeta=0}=0$$

We also noticed that the first non-trivial equation among them corresponds to the value $s=3$ and it is the K.P. equation

(4.3)
$$\left[\Delta_3\Delta_1-\Delta_2\overline{\Delta}_2+d_4\right]\theta(z+\zeta)\theta(z-\zeta)|_{\zeta=0}=0$$

We then wrote down explicitly this equation in the form

(4.4)
$$\begin{aligned}D_1^4\theta\cdot\theta-3D_1^3\theta\cdot\theta+3(D_1^2\theta)^2-3(D_2\theta)^2+3D_1D_3\theta\cdot\theta\\+3D_1\theta\cdot D_3\theta\cdot\theta+d_4\theta\cdot\theta=0\end{aligned}$$

We then noticed that this equation implies the infinitesimal version of Weil decomposition (3.1)

(4.5)
$$\Theta\cap D_1\Theta\subset(D_1^2+D_2)\Theta\cup(D_1^2-D_2)\Theta$$

Now, it was a conjecture of Novikov, later proved by T. Shiota, that *already the first equation among the ones in* (4.2) (*namely the K.P. equation* (4.4)) *suffices to characterize Jacobians. Shiota's theorem* is then the following.

(4.5) *An indecomposable p.p.a.v.* X *is a jacobian if and only if there exist constant vector fields* $D_1\neq, D_2, D_3$ *and a constant* d_4 *s.t.*

$$\left[D_1^4-3D_1D_3+3D_2^2+d_4\right]\theta(z+\zeta)\theta(z-\zeta)\Big|_{\zeta=0}=0$$

This characterization of jacobians is very different in spirit from the preceding ones. In Gunning's criterion one was given a curve $\widetilde{V}_{\alpha,\beta,\gamma}$ to start with. In Welter's criterion one was given the curve $\widetilde{V}_{D_1,D_2}$. In the analytiral formulation of Welters's criterion (cf. (4.1)). The curve is also there; it suffices to consider

$$\zeta(\epsilon) = \sum \epsilon^i W^i \quad ,$$

where $D_i = \sum_{j=1}^{g} W_j^i \frac{\partial}{\partial \zeta_j}$, and once a curve is there one can always try Matsusaka's criterion. In the formulation (4.5) the curve disappears and only its third degree approximation remains, namely D_1, D_2, and D_3.

We would like to give an idea of why theorem (4.5) is true and how one can reconstruct a curve from so little information.

Let us observe that the left hand side of (4.2) only depends on the vector fields $D_1, \ldots, D$ and and the constants $d_4, \ldots, d_{s+1}$ we shall denote the left hand side of (4.2) with the symbo $P_s(D_1, \ldots, D_s, d_4, \ldots, d_{s+1})(z)$ or more simply with $P_s(z)$. Thus in order to prove Novikov' conjecture we must show that

$$(4.6) \qquad \boxed{\begin{array}{l} \exists D_1 \neq 0, D_2, D_3, d_4, \quad \text{s.t.} \quad P_3(z) = 0 \quad \Rightarrow \\ \quad \Rightarrow \exists D_4, D_5, \ldots, d_s, d_{s+1}, \ldots, \quad \text{s.t.} \quad P_s(z) = 0, s \geq 4. \end{array}}$$

In other words one would like to finds D_i 's and d_i 's such that the equations $P_s(z) = 0$, s ≥ 0 are all consequence of $P_3(z) = 0$. Now the equation $P_s\ (z) = 0$ is very complicated. Recal for example, how complicated the equation $P_3(z) = 0$ itself is (cf. (4.4)). But also recall ho simple it becomes when restricted to $\{\theta = D_1\theta = 0\} = D_1\Theta$:

$$P_3(z)|_{D_1\Theta} = (D_1^2 - D_2)\,\theta \cdot (D_1^2 + D_2)\,\theta|_{D_1\Theta}.$$

Let us try to restrict $P_s(z)$ to $D_1\Theta$. Now

$$(4.7) \qquad P_s(z) = \Delta_s\Delta_1 - \Delta_{s-1}\overline{\Delta}_2 + \sum_{i=3}^{s} d_{i+1}\Delta_{s-i}$$

So that the only term in $P_s(z)$ *involving* D_s *and* d_{s+1} *is*

$$2D_1D_s\theta \cdot \theta - 2D_1\theta \cdot D_s\theta + d_{s+1}\theta \cdot \theta$$

and this term vanishes when restricted to $D_1\Theta$. Let us formalize this observation by settir

$$P_s'(z) = P_s\,(D_1, \ldots, D_{s-1}, 0, d_4, \ldots, d_s, 0)$$

The remark we just made con be written as follows:

$$(4.8) \qquad P_s'|_{D_1\Theta} = P_s|_{D_1\Theta}$$

We are now going to prove that

$$(4.9) \qquad P_s'(z)|_{D_1\Theta} = 0 \Leftrightarrow \exists D_s, d_{s+1} \text{ s.t. } P_s(z) = 0$$

So that to solve the equation $P_s(z) = 0$ on X it will suffice to solve it on $D_1\Theta$; in particul we see that the K.P. equation (4.4) is in fact equivalent to the decomposition (4.5). One implication in (4.9) is trivial. To prove the other, first observe that the quasiperiodicity factor of $\theta(z+\zeta)\theta(z-\zeta)$ is an exponential which only depends on z. On the other hand the D_j's are derivations in the ζ-variables, and, using this one easily checks that $P_s(z)$ is a *well defined section* of 2Θ. Now look at the exact sequences

$$0 \to \mathcal{O}_X(\Theta) \xrightarrow{\cdot\theta} \mathcal{O}_X(2\Theta) \to \mathcal{O}_\Theta(2\Theta) \to 0$$

(4.10)

$$0 \to \mathcal{O}_\Theta(\Theta) \xrightarrow{D_1\theta} \mathcal{O}_\Theta(2\Theta) \to \mathcal{O}_{D_1\Theta}(2\Theta) \to 0$$

We are assuming $P_s'(z)|_{D_1\Theta} = 0$. It is well-known that $H^\circ(\Theta, \mathcal{O}_\Theta(\Theta))$ is generated $\frac{\partial\theta}{\partial z_1}, \frac{\partial\theta}{\partial z_1}, \dots \frac{\partial\theta}{\partial z_g}$. Looking at the second exact sequence in (4.10) it follows that there exists a constant vector field D_s such that

$$P_s'(z)|_\Theta = 2D_1\theta \cdot D_s\theta|_\Theta$$

Now look at the section

$$P_s'(z) - 2D_1\theta \cdot D_s\theta + 2D_1D_s\theta \cdot \theta.$$

Its restriction to Θ is zero. Looking at the first exact sequence in (4.10) we find a complex number d_{s+1} such that

$$P_s'(z) - 2D_1\theta \cdot D_s\theta + 2D_1D_s\theta \cdot \theta = -d_{s+1}\theta \cdot \theta$$

proving that there exist D_s and d_{s+1} such that $P_s(z) = 0$.

We can now conclude that, by virtue of (4.9) and by induction, Novikov conjecture (4.6) is equivalent to:

(4.11)
$$\boxed{P_3(z) = \cdots = P_{s-1}(z) = 0 \Rightarrow P_s(z)|_{D_1\Theta} = 0}$$

This statement only involves $D_1, \dots, D_{s-1}, d_4, \dots, d_s$ while (4.9) is in charge of producing D_s and d_{s+1}, *reconstructing the curve by successive approximations* !

Unfortunately we are not able to prove (4.11) in a direct way and, as we do in [AD2], we have to take a technical detour. But we do know how to prove (4.11) under additional hypotheses (cf. [A]). Here for the sake of siplicity we shall make a fairly severe assumption (one which in any case works for a generic curve). Namely we shall assume the $D_1\Theta = \{\theta = D_1\theta = 0\}$ is *reduced.*

Let us define $R_s(z)$ by setting

$$R_s(z) = P_s(z) + \Delta_1 P_{s-1}(z) + \cdots + \Delta_s P_0(z) \quad (P_0 \equiv P_1 \equiv P_2 \equiv 0)$$

Clearly:

$$P_3 = \cdots = P_{s-1} = 0 \Leftrightarrow R_3 = \cdots = R_{s-1} = 0$$

In particular (4.11) is equivalent to

(4.12)
$$\boxed{R_3 = \cdots = R_{s-1} = 0 \Rightarrow R_s|_{D_1\Theta} = 0}$$

A computation gives

$$
\begin{aligned}
R_k = &\Delta_1\widetilde{\Delta}_k\theta\cdot\theta - \widetilde{\Delta}_k\theta\cdot\Delta_1\theta - D_1^2\widetilde{\Delta}_{k-1}\theta\cdot\theta \\
&+ 2\Delta_1\widetilde{\Delta}_{k-1}\theta\cdot D_1\theta - \widetilde{\Delta}_{k-1}\theta\cdot D_1^2\theta \\
&- D_2\widetilde{\Delta}_{k-1}\theta\cdot\theta + \widetilde{\Delta}_{k-1}\theta\cdot D_2\theta + \sum_{i=3}^{k} d_{i+1}\widetilde{\Delta}_{k-i}\theta\cdot\theta
\end{aligned}
$$

where $\widetilde{\Delta}_i = \Delta_i(2D_1,\ldots,2D_i)$. We then get

$$
R_k|_{D_1\Theta} = -\widetilde{\Delta}_{k-1}\theta\cdot(D_1^2 - D_2)\theta|_{D_1\Theta} \tag{4.13}
$$

again a very simple expression. Notice that

$$
\begin{aligned}
R_3|_{D_1\Theta} &= P_3|_{D_1\Theta} \\
&= -\widetilde{\Delta}_2\theta\cdot(D_1^2 - D_2)\theta|_{D_1\theta} \\
&= -(D_1^2 + D_2)\theta\cdot(D_1^2 - D_2)\theta|_{D_1\Theta}
\end{aligned}
$$

and that *by hypothesis* $P_3 = 0$ so that

$$
(D_1^2 + D_2)\theta\cdot(D_1^2 - D_2)\theta|_{D_1\Theta} = 0. \tag{4.14}
$$

In order to prove (4.12) we must show that

$$
\widetilde{\Delta}_{s-1}\theta\cdot(D_1^2 - D_2)\theta|_{D_1\Theta} = 0 \tag{4.15}
$$

Let V be an irreducible component of $D_1\theta$. Since V is reduced, then by (4.14) eith $(D_1^2 - D_2)\theta$ or $(D_1^2 + D_2)\theta$ vanishes on V. In the first case $\widetilde{\Delta}_{s-1}\theta\cdot(D_1^2 - D_2)\theta$ trivial vanishes on V. We may therefore assume that $(D_1^2 - D_2)\theta$ *does not* vanish on V, so th $(D_1^2 + D_2)\theta$ vanishes on V. Since, by induction, $R_{s-1} = 0$ we also have $D_1R_s = 0$. computation gives:

$$
\begin{aligned}
D_1R_{s-1}|_{D_1\Theta} = \Big(&-\widetilde{\Delta}_{s-1}\theta\cdot D_1^2\theta + D_1\widetilde{\Delta}_{s-2}\theta\cdot D_1^2\theta \\
&- \widetilde{\Delta}_{s-2}\theta\cdot D_1^3\theta + \widetilde{\Delta}_{s-2}\theta\cdot D_2\theta \\
&+ \widetilde{\Delta}_{s-2}\theta\cdot D_1D_2\theta\Big)|_{D_1\Theta} = 0
\end{aligned}
$$

The assumption that $(D_1^2 - D_2)\theta|_V \neq 0$ and $R_{s-1} = 0$, implies by (4.13) that $\widetilde{\Delta}_{s-2}\theta|_V =$ Hence (4.16) becomes

$$
\left(-\widetilde{\Delta}_{s-1}\theta\cdot D_1^2\theta + D_1\widetilde{\Delta}_{s-2}\theta\cdot(D_1^2 + D_2)\theta\right)\Big|_V = 0
$$

i.e.

$$
-\widetilde{\Delta}_{s-1}\theta\cdot D_1^2\theta|_V = 0 \tag{4.17}
$$

since $(D_1^2 + D_2)\theta|_V = 0$. Now $(D_1^2 - D_2)\theta|_V \neq 0$ and $(D_1^2 + D_2)\theta|_V = 0$ together gi $D_1^2\theta|_V \neq 0$. So that, by (4.17) we get: $\widetilde{\Delta}_{s-1}\theta|_V = 0$ and hence by (4.13) $R_s|_V = 0$.

Q.E.

. The Kadomtsev Petviashvili hierarchy

In section 3 we saw how the fact that the Riemann theta function of a Jacobian satisfies he trisecant formula (or better its infinitesimal version) could be translated into an infinite et of differential equations satisfied by such a theta function. This set of differential equa- ions is called the *Hirota bilinear form of the Kadomtsev Petviashvili (K.P.) hierarchy*. The Kadomtsev–Petviashvili hierarchy of linear partial differential equations appeared in a rather lifferent context and had its origin in the study of the famous Korteweg–de Vries equation. In his and in the following two sections we shall discuss these connections, essentially following he lines of [SW], [DJKM].

Let us start with the *Korteweg–de Vries* (K.d.V) equation

5.1)
$$2u_{xxx} + 3uu_x - u_t = 0$$

Ve want to rewrite this equation in a different form. Let $D = \frac{d}{dx}$, consider some ring R of unctions of x containing the constants and on which the derivative D is defined and let

$$\mathcal{D} = \left\{ \sum a_i D^i \middle| a_i \in R \right\}$$

e the ring of differential operators with coefficient in R. By inverting D we get the ring *sd* of pseudo differential operators. An element in Psd in an expression of the form

$$\sum_{i=-\infty}^{N} a_i D^i, \qquad a_i \in R$$

nd one has

$$D \cdot D^{-1} = D^{-1} D = 1$$

$$D^{-1} a = \sum_{i=1}^{\infty} (-1)^{i-1} a^{(i-1)} D^{-i}, \qquad a^{(j)} = D^j(a)$$

iven a pseudo – differential operator $L = \sum_{i=-\infty}^{N} a_i D^i$ one sets $L_+ = \sum_{i=0}^{N} a_i D^i$ and alls it the *differential operator part of* L (if $N < \infty, L_+ = 0$). We have, [SW], that if $= D^n + a_{n-2} D^{n-2} + \cdots + a_0$ is a differential operator then there exists a unique n–th root $^{\frac{1}{n}}$ of Q in the ring Psd of the form

$$Q^{\frac{1}{n}} = D + b_1 D^{-1} + b_2 D^{-2} + \ldots$$

particular if

.2)
$$Q = D^2 + u(x,t)$$

e can take

$$Q^{\frac{1}{2}} = D + b_1 D^{-1} + b_2 D^{-2} + \ldots$$

d an easy computation shows that we can write the KdV equation (5.1) in *Lax form*

.3)
$$\frac{\partial Q}{\partial t} = \left[Q_+^{3/2}, Q \right].$$

his allows us to generalize the KdV equation in various directions. First of all we can assume at the function u in (5.2) depends on infinitely many variables

$$u = u(t_1, t_3, t_5, \dots) \quad t_1 = x,\ t_3 = t$$

and we can write

$$\frac{\partial Q}{\partial t_i} = \left[Q_+^{i/2}, Q\right]. \tag{5.4}$$

This is the so called *KdV hierarchy* for the operator Q. Secondly we can directly start from a pseudo–differential operator

$$L = D + \sum_{i=-1}^{\infty} u_i D^{-i}, \quad u_i = u_i(t_1, t_2, \dots) \tag{5.5}$$

and write the equations

$$\frac{\partial L}{\partial t_i} = \left[L_+^i, L\right]. \tag{5.6}$$

This is the so called *Kadomtsev–Petviashvili hierarchy* (KP hierarchy). It is not hard to see that, if $L^2 = Q = D^2 + u$, then L is a solution of (5.6) if and only if Q is a solution of (5.4). Clearly we can explicitly write (5.6) as a set of infinitely many complicated non–linear partial differential equations for the coefficients of L. In particular using (5.6) for $i = 2$, and setting $u = u_1$, $x = t_1$, $y = t_2$, $t = t_3$ we get the following equation for u

$$\frac{\partial}{\partial x}(2u_{xxx} - 3uu_x - u_t) + 3u_{yy} = 0$$

which is the *KP equation for* u.

We are now going to explain a systematic way, originally due to Sato, to construct solutions of (5.6). First of all we want to rephrase our original equations in terms of eigenfunctions for L. For this let us introduce the space M consisting of all "functions" of the form

$$e^{\sum_{i\geq 1} t_i z^{-i}} f(z,t), \qquad t = (t_1, t_2, \dots)$$

$$f(z,t) = \sum_{i=-N}^{\infty} a_i(t) z^i, \qquad a_i(t) \in R$$

we give M the structure of a module over *Psd* by setting

$$a e^{\sum t_i z^{-1}} f(z,t) = e^{\sum t_i z^{-i}} a f(z,t), \text{ if } a \in R$$

$$D\left(e^{\sum t_i z^{-i}} f(z,t)\right) = e^{\sum t_i z^{-i}}\left(z^{-1} f(z,t) + D(f)\right),\ D = \frac{d}{dx} = \frac{\partial}{\partial t_1}$$

$$D^{-1}\left(e^{\sum t_i z^{-i}} f(z,t)\right) = e^{\sum t_i z^{-i}}\left(\sum_{i=-N}^{\infty}\left(\sum_{h=1}^{\infty}(-1)^{h-1} a_i^{(h-1)} z^h\right) z^i\right)$$

It is easy to see that M is a free cyclic module generated by $e^{\sum t_i z^{-i}} \cdot z^{-1}$ and that given $Q = \sum_{i=-\infty}^{N} a_i D^i$ one has

$$Q e^{\sum t_i z^{-i}} \cdot z^{-1} = e^{\sum t^i z^{-i}} \sum_{i=-N}^{\infty} a_i z^{i-1}.$$

Now let L be as in (5.5). It is easy to see that there exists an operator $K = 1 + \sum_{i=1}^{\infty} a_i D^{-i}$

with the property that

$$L\,K = K\,D \text{ or equivalently}$$

(5.7)
$$L = K\,DK^{-1}$$

Consider now the element of M defined as follows

$$\psi(z,t) = Ke^{\Sigma t_i z^{-i}}$$

(5.8)
$$e^{\Sigma t_i z^{-i}}\left(z^{-1} + \sum_{i=1}^{\infty} a_i z^{i-1}\right)$$

We shall call ψ a *wave function for* L. We have that

$$\begin{aligned} L\psi(z,t) &= K\,DK^{-1}\left(Ke^{\Sigma t_i z^{-i}} z^{-1}\right) \\ &= K\,D\left(e^{\Sigma t_i z^{-i}} z^{-1}\right) \\ &= K\,z^{-1}\left(e^{\Sigma t_i z^{-i}} z^{-1}\right) = z^{-1}\psi(z,t) \end{aligned}$$

Thus $\psi(z,t)$ is an eigenfunction for L with eigenvalue z^{-1}. Since K is clearly determined only up to right multiplication by *constant coefficient operators* of the form $1+\sum_{i\geq 1} b_i D^{-i}$, the eigenfunction ψ is not uniquely determined by L. On the other hand given $\psi = e^{\Sigma t_i z^{-i}}\left(z^{-1}+\sum_{i=1}^{\infty} a_i z^i\right)$ if we set $K = 1+\sum_{i=1}^{\infty} a_i D^{-i}$ and $L = KDK^{-1}$ we get that ψ is a wave function for L. The following Proposition allows us, in order to find a solution of (5.6), to check a bilinear relation for a wave function of L and another function of the form $e^{\Sigma t_i z^{-i}}\left(z^{-1}+\sum_{i=1}^{\infty} b_i z^i\right)$.

(5.9) **Proposition** *Let L be as in (5.5) and let ψ be a wave function for L. Then L is a solution for the KP hierarchy if and only if there exists $\psi^{\#} = e^{\Sigma t_i z^{-i}}\left(z^{-1}+\sum_{i=1}^{\infty} b_i(t) z^i\right)$ such that*

(5.10)
$$\mathrm{Res}_{z=0}\psi(z,t)\psi^{\#}(z,t')dz = 0$$

Proof. We shall only prove that if there exists $\psi^{\#}$ as above such that (5.10) is satisfied then L is a solution of the KP hierarchy. Assume for the moment that we have shown that (5.10) implies that for any $r\geq 1$

(5.11)
$$\frac{\partial\psi}{\partial t_r} = L_+^r\psi.$$

Notice that since M is a free cyclic module and $\psi\neq 0$ it suffices to show that

$$\left(\frac{\partial L}{\partial t_r} - [L_+^r, L]\right)\psi = 0$$

If $S_r = \frac{\partial K}{\partial t_r}K^{-1}$, where $\psi(z,t) = Ke^{\Sigma t_i z^{-i}} z^{-1}$, we then have

(5.12)
$$\frac{\partial L}{\partial t_r} = \frac{\partial K}{\partial t_r} D\,K^{-1} - K\,D\,K^{-1}\frac{\partial K}{\partial t_r}K^{-1} = [S_r, L]$$

On the other hand

$$\frac{\partial\psi}{\partial t_r} = \frac{\partial}{\partial t_r} K e^{\Sigma t_i z^{-i}} z^{-1} =$$
$$= \frac{\partial K}{\partial t_r} e^{\Sigma t_i z^{-i}} z^{-1} + z^{-r}\psi$$
$$= S_r\psi + L^r\psi$$

So

$$[L^r_+, L]\,\psi = L^r_+ L\psi - LL^r_+\psi$$
$$= z^{-1}\frac{\partial\psi}{\partial t_r} - L\frac{\partial\psi}{\partial t_r}$$
$$= z^{-1}S_r\psi + z^{-1}L^r\psi - LS_r\psi - L^{r+1}\psi$$
$$= S_r L\psi + L^{r+1}\psi - LS_r\psi - L^{r+1}\psi$$
$$= [S_r, L]\,\psi$$

Comparing with (5.12) this proves our claim. It thus remains to show that (5.10) impli (5.11). An easy direct computation shows that

$$\widehat{\psi} = \frac{\partial\psi}{\partial t_r} - L^r_+\psi$$
$$= e^{\Sigma t_i z^{-i}}\left(\sum_{i=0}^{\infty} c_i z^i\right)$$

Now, acting with the operator $\frac{\partial}{\partial t_r} - L^r_+$ on both sides of (5.10) we get

$$\mathrm{Res}_{z=0}\widehat{\psi}(z,t)\psi^{\#}(z,t')dz = 0,$$

so our claim follows from the

(5.13) **Lemma** *Let* $\mu = e^{\Sigma t_i z^{-i}}\left(\sum_{j=0}^{\infty} c_j z^j\right)$ *and* $\nu = e^{\Sigma t_i z^{-i}}\left(z^{-1} + \sum_{i=1}^{\infty} b_i z^{i-1}\right)$. *If*

$$Res_{z=0}\mu(z,t)\nu(z,t')dz = 0$$

then $\mu(z,t) = 0$.

Proof. We prove that $c_j = 0, \forall j \geq 0$ by induction on j. Letting $t = t'$ we get

$$0 = \mathrm{Res}_{z=0}\mu(z,t)\nu(z,t)dz = c_0$$

Suppose now that $c_0 = c_1 = \cdots = c_{j-1} = 0$. Differentiating with respect to t_j and setti $t = t'$ we get

$$0 = \mathrm{Res}_{z=0}\frac{\partial\mu}{\partial t_j}(z,t)\nu(z,t) = c_j,$$

proving our claim.

The above Proposition shows that, in order to find solutions of (5.6), it suffices to prod a pair of vave functions ψ and $\psi^{\#}$ as in (5.8) satisfying the bilinear relation (5.10). This what we are going to do next.

. The infinite dimensional Grassmannian

Let $\mathbf{C}$ be the complex plane, set

$$D_\varepsilon = \{z \in \mathbf{C} : |z| < \varepsilon\}$$
$$\dot{D}_\varepsilon = \{z \in \mathbf{C} : 0 < |z| < \varepsilon\}$$

Ve denote by $\mathcal{O}(D_\varepsilon), \left(\mathcal{O}\left(\dot{D}_\varepsilon\right)\right)$, the space of holomorphic functions on D_ε, (resp. D_ε). Ve set

$$H = \lim_{\varepsilon \to 0} \mathcal{O}(D_\varepsilon)$$

Ve let

$$H_+ = \lim_{\varepsilon \to 0} \mathcal{O}(D_\varepsilon) = \mathbf{C}\{z\}$$

nd denoting by $\widehat{\mathbf{C}}$ the Riemann sphere obtained as $\widehat{\mathbf{C}} = \mathbf{C} \cup \{\infty\}$ we set

$$H_- = \{f \text{ holomorphic in } \widehat{\mathbf{C}} - \{0\}, \text{ with } f(\infty) = 0\}$$

hen it is easy to see that

$$H = H_- \oplus H_+$$

'ix $\varepsilon > 0$. Consider the spaces

$$\mathcal{H}_-^\varepsilon = \left\{f : f \in L^2(\partial D_\varepsilon), f \text{ holo on } \widehat{\mathbf{C}} - \overline{D}_\varepsilon,\ f(\infty) = 0\right\}$$
$$\mathcal{O}\left(\overline{D}_\varepsilon\right) = \{f \in \mathcal{O}(D_\varepsilon) : f \text{ bounded on } \partial D_\varepsilon\}$$

ıen $\mathcal{H}_-^\varepsilon$ is a Hilbert space, $\mathcal{O}\left(\overline{D}_\varepsilon\right)$ is a Banach space, $H_- = \varprojlim \mathcal{H}_-^\varepsilon$ is a Fréchet space, nd $H_+ = \lim_{\varepsilon \to 0} \mathcal{O}\left(\overline{D}_\varepsilon\right)$ has the direct limit topology. Finally we give H the product ›pology. We shall denote by $\pi_+ : H \longrightarrow H_+$, $\pi_- : H \longrightarrow H_-$ the obvious projections. 'iven $f, g \in H$ we define the pairing

$$(f, g) = \operatorname{Res}_{z=0} f g dz \tag{.1}$$

ı particular we get a perfect pairing

$$H_+ \times H_- \longrightarrow \mathbf{C}$$

ıd we get that H_+ and H_- are topological dual of each other and that H coincides with s topological dual. Consider now the spaces $M(H_+, H_-)$ and $M(H_-, H_+)$ of bounded ıear operators from H_+ to H_- and from H_- to H_+ respectively. One can show that ven $\varphi \in M(H_+, H_-)$, $(\varphi \in M(H_-, H_+))$, there exists a unique function $K_\varphi(z, w)$ ›lomorphic on $\widehat{\mathbf{C}} \times \widehat{\mathbf{C}} - \{0\} \times \widehat{\mathbf{C}} \cup \widehat{\mathbf{C}} \times \{0\}$ and such that $K_\varphi(\infty, w) = K_\varphi(z, \infty) = 0$, esp. $K_\varphi(z, w) \in \mathbf{C}\{z, w\}$), such that

$$\varphi(f(z)) = \operatorname{Res}_{z=0} K_\varphi(z, w) f(w) dw$$

particular one sees that given $\varphi \in M(H_+, H_-)$, $\psi \in M(H_-, H_+)$ one has a *well defined* ace for $\varphi\psi$ and $\psi\varphi$ given by

$$\begin{aligned} Tr(\varphi\psi) &= Tr(\psi\varphi) \\ &= \operatorname{Res}_{\substack{z=0 \\ w=0}} K_\varphi(z, w) K_\psi(z, w) dz\, dw \end{aligned} \tag{.2}$$

we shall use this later. For now notice that given $\varphi \in M(H, H)$, we can think of φ as a tw by two matrix of operators

$$\varphi = \begin{pmatrix} \varphi_{++} & \varphi_{+-} \\ \varphi_{-+} & \varphi_{--} \end{pmatrix}$$

where

$$\varphi_{++} = \pi_+ \varphi \big|_{H_+}, \ \varphi_{+-} = \pi_+ \varphi \big|_{H_-}, \ \varphi_{-+} = \pi \varphi \big|_{H_+}, \ \varphi_{--} = \pi_- \varphi \big|_{H_-}$$

and using (6.2) we can define a 2–cocycle on $M(H, H) = \mathbf{g}\,\mathbf{l}(H)$ considered as a lie algebr by

(6.3) $$c(\varphi, \psi) = tr\,(\varphi_{+-}\psi_{-+} - \psi_{+-}\varphi_{-+})$$

let us finally introduce the group

$$G = \{\varphi \in M(H, H) : \varphi_{--} \text{ is Fredholm of index } 0\}.$$

We now define our indexing set for solution of the KP hierarchy.

(6.4) **Definition** *The infinite grassmannian $Gr(H)$ is the collection of closed subspaces W H such that*

$$\pi_- : W \longrightarrow H_-$$

is a Fredholm operator of index 0.

On as

(6.5) **Proposition** *$Gr(H)$ is an infinite dimensional complex manifold based on the spa $M(H_-, H_+)$.*

Proof. We shall explain how to give a system of charts for $Gr(H)$ leaving the details to t reader.

Fix a sequence of integers $s = (h_1, h_2, \ldots)$ with $h_i > h_{i+1}$ and $h_i = -i, \quad i \gg 1$.

Let $H_-^s = \{f = \sum a_i z^i : a_i = 0 \text{ if } i \in s\}$ and $H_+^s = \{f = \sum a_i z^i : a_i = 0 \text{ if } i \notin s\}$. Th it is easy to see that $H_-^s \in Gr(H)$ and $H_-^s \cong H_-$ by the isomorphism

(6.6) $$\sum_{i=1}^{\infty} a_i z^{h_i} \longrightarrow \sum_{i=1}^{\infty} a_i z^{-i}$$

similarly $H_+^s \cong H_+$. Notice now that $H = H_-^s \oplus H_+^s$. Using the above isomorphisms we g an isomorphism

$$M(H_-, H_+) \cong M(H_-^s, H_+^s)$$

and given $\varphi \in M(H_-^s, H_+^s)$ we can consider the space

$$W_\varphi = \{f + \varphi(f) : f \in H_-^s\}$$

It is easy to see that $W_\varphi \in Gr(H)$. The collection of these spaces gives us a chart C_s $Gr(H)$, $C_s \cong M(H_-^s, H_+^s) \cong M(H_-, H_+)$. Now one has to show that $Gr(H) = \bigcup_s C_s$, i that given $W \in Gr(H)$ then exists a sequence $s = (h_1, h_2, \ldots)$ such that, denoting by π $H \longrightarrow H_-^s$ the projection with kernel H_+^s, one has that $\pi_-^s : W \longrightarrow H_-^s$ is an isomorphis For this consider the spaces $H_n = \left\{\sum_{i \geq n} a_i z^i\right\}$, $n \in \mathbf{Z}$ and set $W_n = H_n \cap W$. Since ssumption $W_0 = W \cap H_+$ is finite dimensional it follows that, for $n \gg 0$, $W_n = \{0\}$. We nen define the sequence $s = (h_1 > h_2 > \ldots)$ by $h \in s$ if $W_n \supseteq W_{n-1}$. We leave to the eader to verify that $W \in C_s$.

In particular, in $Gr(H)$, we have the chart C_{s_0} where $s_0 = (-1, -2, \ldots)$ i.e. the set f spaces W such that $\pi_- : W \longrightarrow H_-$ is an isomorphism. We shall call these spaces *ansversal.* We now remark that the group G acts transitively on $Gr(H)$. To see this we let $\in Gr(H)$ and show that there exists $g \in G$ such that $gH_- = W$. Assume first $W = H_-^s$ r some $s = (h_1, h_2, \ldots)$, $h_i = -i$, $i \gg 1$. Let $f \in H$ and write $f = f_+ + f_-$, $f_\pm \in H_\pm$,

$$f_+ = \sum_{i \geq 0} a_i z^i \quad , \quad f_- = \sum_{i<0} a_i z^i .$$

et $\mathbf{Z} - s = \{k_0, k_1, \ldots\}$ with $k_i = i$, $i \gg 0$. Set $g(f) = g(f_+) + g(f_-) = \sum_{i\geq 0} a_i z^{k_i} +$ $_{i<0} a_i z^{h_i}$. It is then clear that $g \in G$. For a general $W \in Gr(H)$ assume that $W \in C_s$ r some $s = (h_1, h_2, \ldots)$. Then W is the graph of an operator $\gamma : H_-^s \to H_+^s$. Since $= H_-^s \oplus H_+^s$ let g be the operator defined by $g\Big|_{H_+^s} = id$, $g\Big|_{H_-^s} = id + \gamma$. It is then sy to verify that $g \in G$ and $W = gH_-^s$. This proves that W can be obtained by an ement of G starting from H_-^s. From what we have seen it suffices to show that given $\in Gr(H)$, $g \in G$, then $gW \in Gr(H)$. Writing $W = hH$, $h \in G$ we can see that it ffices to see $(gh)H_- \in Gr(H)$. This is true by the very definition of G.

For a while we shall be interested in the action of a particular subgroup of G, namely the bgroup Γ_- of nowhere vanishing functions $g(z)$ on $\widehat{\mathbb{C}} - \{0\}$ such that $g(\infty) = 1$, acting on as multiplication operator. Notice that since such g maps H_- isomorphically into itself - is contained in G. Also given $g \in \Gamma_-$, since $\widehat{\mathbb{C}} - \{0\}$ is simply connected and $g(\infty) = 1$ e have

$$g = e^f, \quad f \in H_-$$

exp : $H_- \longrightarrow \Gamma_-$ is a group isomorphism.

Now fix $W \in Gr(H)$ and assume that there exists $f \in H_-$ such that $e^f W$ is transversal his is always true as we shall see at the end of section 7.). Let

$$H_-^W = \{f \in H_-, \ e^{-f} W \text{ is transversal}\} .$$

w, as we just said, $H_-^W \subset H_-$ is a non-empty open set. Then for any $f \in H_-^W$ set

.7) $$\widetilde{\psi}_W(z, f) = \pi_-^{-1}(z^{-1}) \cap (e^{-f} W)$$

$\widetilde{\psi}_W(z, f)$ is the unique element in $e^{-f} W$ such that $\pi_-\left(\widetilde{\psi}_W(z, f)\right) = z^{-1}$. Let

$$\psi_W(z, f) = e^f \widetilde{\psi}_W(z, f) \in W.$$

tice that if we write $f = \sum t_i z^{-i}$

8) $$\psi_W(z, f) = e^{\sum t_i z^{-i}} \left(z^{-1} + \sum_{i \geq 1} a_i(t) z^{i-1} \right)$$

f the form (5.8). We shall see later that the $a_i(t)$ are meromorphic functions on H_-.

We now many ask if $\psi_W(z, t)$ gives a solution to the KP equation i.e. if we can find a

$\psi^{\#}(z,\ t)$ of the same form such that

(6.9) $$\mathrm{Res}_{z=0}\ \psi\ (z,\ t)\psi^{\#}(z,\ t')dz = 0$$

To see this let $W^{\perp} = \{g \in H :\ \mathrm{Res}_{z=0}\ g\tilde{f}dz = 0,\quad \forall \tilde{f} \in W\}$.

(6.10) **Proposition** $W^{\perp} \in Gr(H)$.

Proof. Consider the operator
$$\varphi : W \oplus H_{+} \longrightarrow H$$
given by $\varphi(f,\ f_{+}) = f + f_{+},\ f \in W,\ f_{+} \in H_{+}$. By assumption φ is Fredholm of index and so is the induced operator
$$\tilde{\varphi} : H_{+} \longrightarrow H/W.$$
Now it is easy to see that $(H/W)^{*} = W^{\perp}$. Thus since $H_{-} = H_{+}^{*}$, taking the adjoint operatc to $\tilde{\varphi}$ we get that
$$\pi_{-} : W^{\perp} \longrightarrow H_{-}$$
is Fredholm of index 0.

By the above Proposition if we set $\psi^{\#}(z,t) = \psi_{W^{\perp}}$ we immediately set that (6.9) i verified and hence we can associate to W a solution of the KP hierarchy.

We finally notice the following

(6.11) **Corollary** *Let W in $Gr(H)$, $K_W = Ker\ (\pi_{-} : W \longrightarrow H_{-})$, $C_W = Coker\ (\pi_{-}$ $W \longrightarrow H_{-})$, then $K_W = C^{*}_{W^{\perp}}$, $C_W = K^{*}_{W^{\perp}}$.*

Proof. Clear from the proof of Proposition (6.10).

7. The τ function

In this section we shall construct, following [SW], a line bundle on $Gr(H)$ and introdu the so called *τ-function.*

(7.1) **Definition.** *Let $\alpha : H_{-} \to H$ be an operator. We say that ψ is admissible if $(\pi_{-}\alpha - I$ $H_{-} \to H_{+}$ lies in the ideal of $M(H_{-}, H_{-})$ generated by operators of the form uv whe $u : H_{+} \to H_{-}$, $v : H_{-} \to H_{+}$. If $W \in Gr(H)$ and if $\alpha : H_{-} \to W$ is an isomorphis such that $i\alpha : H_{-} \to H$, i being the inclusion, is admissible we say that α is an admissil isomorphism.*

(7.2) **Proposition.** *Let $\alpha : H_{-} \to H$ be an admissible operator, then setting $a = \pi_{-}\alpha -$ the trace $Tr(\wedge^{i}a)$ is well defined for all $i \geq 1$, the operator $\wedge^{i}a$ being the one induced exterior powers. Furthermore the series*
$$\det(\alpha) = 1 + \sum_{i \geq 1} Tr(\wedge^{i}a)$$
converges.

Proof. We assume for simplicity that $a = uv$ with $v : H_- \to H_+$, $u : H_+ \to H_-$. Then we have already seen that $Tr(a)$ is well defined and similarly one sees that $Tr(\wedge^i a)$ is well defined also. It remains to show that $\det(\pi_- \alpha - I)$ is well defined.

Let $K_v(K_w)$, be the kernel of v (resp. u). Then there exists $\varepsilon > 0$ such that $K_v(z, w)$ converges for $|z| < \varepsilon, |w| < \varepsilon$. Let $\mathcal{H}_-^{\varepsilon/2}$, (resp. $\mathcal{H}_+^{\varepsilon/2}$), be the Hilbert space of L^2 functions on $\partial D_{\varepsilon/2}$ which extend to holomorphic functions on $\widehat{\mathbb{C}} - \overline{D}_{\varepsilon/2}$ (resp. on $D_{\varepsilon/2}$) and have value 0 at infinity. We have obvious operators

$$H_- \xrightarrow{j} \mathcal{H}_-^{\varepsilon/2}, \quad \mathcal{H}_+^{\varepsilon/2} \xrightarrow{p} H_+$$

and by the property of v we clearly get an operator $\widetilde{v} : \mathcal{H}_-^{\varepsilon/2} \to \mathcal{H}_+^{\varepsilon/2}$ such that

(7.3) $$a = uv = up\widetilde{v}j.$$

Let $\widetilde{a} = jup\,\widetilde{v}$. It is easy to see that $\widetilde{a}$ is a trace class operator and $Tr(\widetilde{a}) = Tr(a)$. A similar argument shows that $Tr(\wedge^i \widetilde{a}) = Tr \wedge^i (a)$. Hence $1 + \sum_{i \geq 1} (\wedge^i a) = 1 + \sum_{i \geq 1} (\wedge^i \widetilde{a})$ and the series on the right hand side converges.

We shall denote by $A(H_-)$ the group of admissible isomorphism of H_-, and by $A(W)$ the space of admissible isomorphism $\psi : H_- \to W$, $W \in Gr(H)$

(7.4) **Lemma** *Let $W \in Gr(H)$ then $A(W) \neq \phi$ and it is a principal homogeneous space for $A(H_-)$ under the action which associates to $\psi \in A(W)$ and $h \in A(H_-)$ the isomorphism ψh^{-1}.*

Proof. We have seen that for any $W \in Gr(H)$ there exists a sequence $s = (h_1 > h_2 > \ldots)$, $h_i = -i$, for $i \gg 1$, such that the projection $\pi_-^s : W \to H_-^s$ is an isomorphism. Let $j^s : H_-^s \to H$ be the isomorphism defined in (6.6). Then $(j^s \pi_-^s)^{-1} : H_- \to W$ is clearly admissible. The rest of the lemma is clear.

The preceding lemma allow us to construct a line bundle on $Gr(H)$. Consider the space

(7.5) $$\mathcal{S} = \{(W, \alpha, \lambda) : W \in Gr(H), \alpha \in A(W), \lambda \in \mathbb{C}\}$$

We define an equivalence relation of $\mathcal{S}$ as follows

(7.6) $$(W, \alpha, \lambda) \sim (W', \alpha', \lambda') \quad \text{if}$$

$$W = W', \quad \lambda = \lambda' \det(\alpha'^{-1} \alpha)$$

Notice that by (7.4) $\alpha'^{-1}\alpha \in A(H_-)$ so that by (7.2) our definition makes sense. We let $\mathcal{L} = \mathcal{S}/\sim$. By (7.4) it is clear that this is a holomorphic line bundle on $Gr(H)$, (this is called Det^{-1} in [SW]).

We now define a section of $\mathcal{L}$ as follows. Given $W \in Gr(H)$, let $\alpha \in A(W)$ we set

(7.7) $$\delta(W) = (W, \alpha, \det(\pi_- \alpha)), \quad \text{mod} \sim.$$

It is clear that the element defined above is independent of the choice of α and defines a holomorphic section of $\mathcal{L}$. Notice that $\delta(W) \neq 0$ if and only if W is transversal. We want to give an alternative description of the fiber $\mathcal{L}_W$, for $W \in Gr(H)$.

(7.8) **Proposition.** *Let $W \in Gr(H)$; let K_W and C_W be as in (6.11), then there is a* *canonical isomorphism*

$$\mathcal{L}_W \cong \operatorname{Hom}\left(\bigwedge^{\max} K_W, \bigwedge^{\max} C_W\right)$$

Proof. To get our proposition it suffices to give a canonical non-degenerate pairing betwee $\mathcal{L}^*_W$ and $M_W = \operatorname{Hom}\,(\bigwedge^{\max} K_W, \bigwedge^{\max} C_W)$. Notice that

$$\mathcal{L}^*_W = \{(\alpha, \lambda) : \alpha \in A(W), \lambda \in \mathbf{C}\}/\sim$$

where $(\alpha, \lambda) \sim (\alpha', \lambda')$ if $\lambda' = \lambda \det({\alpha'}^{-1}\alpha)$. We now choose in W a closed subspac complementary to K_W (this is possible since K_W is finite dimensional) and in H_- a spac C' complementary to $\pi(W)$, so that the restriction of the projection $H_- \to C_W$ is a isomorphism when restricted to C'. Finally we choose a fixed isomorphism $\varphi_0 : K_W \to C_W$ so that we also get an isomorphism $\varphi_0' : K_W \to C'$. Having made these choices we get a isomorphism

$$\pi'_- : W \to H_-$$

defined as follows. If $w \in W$, we write $w = w' + k, \quad w' \in W', \quad k \in K_W$, then we set

$$\pi'_-(w) = \pi_-(w') + \varphi_0'(k). \tag{7.9}$$

Suppose now given $l \in \mathcal{L}^*$, and $m \in \operatorname{Hom}\,(\bigwedge^{\max} K_W, \bigwedge^{\max} C_W)$. Let (α, λ) be a representativ for l, $\quad m = \mu \overset{max}{\wedge} \varphi_0, \quad \mu \in \mathbf{C}$.
We set

$$< l, m >= \mu\lambda \det(\pi'_-\alpha). \tag{7.10}$$

Notice that, since $\pi'_-\alpha$ differs from $\pi'_-\alpha$ by a finite rank operator, $\pi'_-\alpha$ is admissible an so (7.10) makes sense. It is now a tedious but straight-forward exercise to check that th pairing (7.10) does not depend from the choices made so this gives our claim.

A useful corollary of the above proposition is the following

(7.11) **Corollary.** *Let $\mu : Gr(H) \to Gr(H)$ be the involution defined by $\mu(W) = W^\perp$. Th $\mu^*(\mathcal{L})$ is canonically isomorphic to $\mathcal{L}$ and furthermore $\mu^*\sigma(W) = \sigma(W^\perp)$.*

Proof. For the first claim one has just to put together proposition (7.8) with corollary (6.1 As for the second notice that W is transversal if and only if $W^\perp$ is, so that $\mu^*\sigma(W)$ h the same zero set as σ. Since one knows (see [SW] for a proof which applies verbatim to o case) that there are no non-constant holomorphic functions on $Gr(H)$ we get that $\mu^*(\sigma)$ is non-zero multiple of σ. To see that they are equal it suffices to see that $\mu^*\sigma(H_-) = \sigma(H_-$ We leave this verification to the reader.

We can restate (7.11) by saying that there exists a map $\widetilde{\mu} : \mathcal{L} \to \mathcal{L}$ such that

$$\begin{array}{ccc} \mathcal{L} & \xrightarrow{\ \widetilde{\mu}\ } & \mathcal{L} \\ \downarrow & & \downarrow \\ Gr(H) & \xrightarrow{\ \widetilde{\mu}\ } & Gr(H) \end{array} \tag{7.12}$$

ommutes, (the vertical arrows being the projection). The map μ is linear on the fibres and $^2 = id$. Furthermore $\tilde{\mu}(\sigma(W)) = \sigma(W^{\perp})$. We have seen in the previous paragraph that $r(H)$ is a homogeneous space under the group G of isomorphism

$$\varphi = \begin{pmatrix} \varphi_{++} & \varphi_{+-} \\ \varphi_{-+} & \varphi_{--} \end{pmatrix}$$

 H such that φ_{--} is Fredholm of index 0. We would like to define a compatible action G on $\mathcal{L}$. This is not possible but we can see instead that we can define an action of a uitable non-trivial central extension

.13) $$1 \to \mathbf{C}^* \to \widehat{G} \to G \to 1$$

 see this we set

$$\begin{aligned} \widetilde{G} &= \{(g, a) : g \in G, \quad a \in Gl(H_-), \quad g_{--}\, a^{-1} \text{ is admissible}\} \\ &\subseteq G \times Gl(H_-) \end{aligned}$$

 is clear that $\widetilde{G}$ is a subgroup of $G \times Gl(H)$.
e set

$$K = \{(1, a) : \det a = 1\}$$

nd

.14) $$\widehat{G} = G/K$$

 order to give an action of $\widehat{G}$ on $\mathcal{L}$ it then suffices to define an action of $\widehat{G}$ on $\mathcal{S}$ and verify rious compatibilities. Given $(g, a) \in \widehat{G}$, and $(W, a, \lambda) \in \mathcal{S}$ we set

$$(g, a) \cdot (W, a, \lambda) = (g\,W, g\alpha\, a^{-1}, \lambda).$$

 is a trivial matter to see that $(g\,W, g\alpha\, a^{-1}, \lambda) \in \mathcal{S}$ and that this action induces an action $\widehat{G}$ on $\mathcal{L}$. In section 6 we considered the subgroup $\Gamma_- = \exp(H_-) \subset G$; notice that if $\in \Gamma_-$, then $g_{--} = a : H_- \to H_-$ is an isomorphism, so that we can lift the inclusion $_- \subset G$ to an inclusion $\Gamma_- \subset \widehat{G}$ by sending $g \in \Gamma_-$ to the equivalence class of the pair $, a)$. From now on we shall consider Γ_- as a subgroup of $\widehat{G}$. In particular Γ_- acts on $\mathcal{L}$. e can thus give the following

.15) **Definition** *Let* $l \in \mathcal{L}_W^* = \mathcal{L}_W - \{0\}$ *we set*

$$\tau_l(f) = \frac{\sigma(e^{-f}W)}{e^{-f}l}, \quad \forall f \in H_-.$$

We can consider $\tau_l(f)$ as a holomorphic function on $\mathcal{L}^* \times H_-(\mathcal{L}^* = \mathcal{L}$-zero section) or, a fixed l, as a holomorphic function on H_-. Notice also that if $l' = sl, \quad s \in \mathbf{C}^*$ then

$$\tau_{l'}(f) = \alpha^{-1}\tau_l(f),$$

 that, up to multiplication by a non-zero scalar, $\tau_l(f)$ depends only on $W \quad (l \in \mathcal{L}_W^*)$. view of this we shall sometimes improperly write $\tau_W(f)$ for $\tau_l(f)$. In particular if W is nsversal we can take $l = \sigma(W)$ and normalize the τ-function as follows

$$\tau_W(f) = \frac{\sigma(e^{-f}W)}{e^{-f}\sigma(W)}.$$

Given $W \in Gr(H)$ we choose an admissible isomorphism $\alpha : H_- \to W$ and a point $l \in \mathcal{L}^*_W$ represented by $(\alpha, 1)$ and we set $\alpha_- = \pi_-\alpha, \quad \alpha_+ = \pi_+\alpha$ and $g = e^{-f}$, then we get

$$\tau_l(f) = \det\left(\alpha_- + g_{--}^{-1} g_{-+} \alpha_+\right) \tag{7.16}$$

In particular if W is transversal and normalized as above

$$\tau_W(f) = \det\left(I + g_{--}^{-1} g_{-+} \alpha_2\right) \tag{7.17}$$

whenever W is the graph of the operator $\alpha_2 : H_- \to H_+$

We wish now to express $\tau_{W^\perp}$ in terms of τ_W. We have

(7.18) **Proposition.** *Let* $l \in \mathcal{L}^*_W$ *then*

$$\tau_l(f) = \tau_{\tilde{\mu} l}(-f).$$

Proof. Notice that under our pairing the operator e^f is self-adjoint. It follows that

$$\mu(e^f W) = (e^f W)^\perp = e^{-f} W^\perp = e^{-f}\mu(W).$$

We want to see that also

$$\tilde{\mu}(e^f l) = e^{-f}\tilde{\mu}(l), \quad \forall\, l \in \mathcal{L}^*.$$

By (7.11) we have that $\tilde{\mu}\, e^f \tilde{\mu}\, e^f$ is an automorphism of $\mathcal{L}$, hence the multiplication by non-zero scalar. Computing on H_- one immediately shows that this scalar is equal to 1. B definition

$$\begin{aligned}
\tau_l(f) &= \frac{\sigma(e^{-f}W)}{e^{-f}l} \\
&= \frac{\tilde{\mu}\,\sigma(e^{-f}W)}{\tilde{\mu}(e^{-f}l)} \\
&= \frac{\sigma(e^f W^\perp)}{e^f \tilde{\mu} l} \\
&= \tau_{\tilde{\mu} l}(-f).
\end{aligned}$$

In what follows we shall assume for simplicity that W is transversal but the reader, he whishes, will have no difficulty in extending our discussion to the general case. Given transversal $W \in Gr(H)$ such that W is the graph of $\alpha : H_- \to H_+$, and given $g = e^f$, have seen that

$$\tau_W(f) = \det(I + g_{--}^{-1} g_{-+} \alpha)$$

We want to extend this function to a slightly larger space than H_-. For this let $K(z, w)$ the kernel of α. There exists $\varepsilon > 0$ such that $K(z, w)$ converges $|z| < 2\varepsilon, |w| < 2\varepsilon$, and have already seen that there is

$$\tilde{\alpha} : \mathcal{H}^\varepsilon_- \to \mathcal{H}^\varepsilon_+$$

uch that $\alpha = f\tilde{\alpha}i$. Notice now that $\mathcal{H}^\epsilon = \mathcal{H}^\epsilon_- \oplus \mathcal{H}^\epsilon_+ = L^2(\partial D_\epsilon)$, so if we take $f \in \mathcal{H}^{\epsilon/2}_-$ learly $g = e^{-f}$ acts on $\mathcal{H}^\epsilon$ as a multiplication operator, and sends $\mathcal{H}^\epsilon_-$ isomorphically to tself. Defining g_{--} and g_{-+} in the obvious way we also see that $g_{--}g_{-+}\tilde{\alpha} : \mathcal{H}^\epsilon_- \to \mathcal{H}^\epsilon$ is trace class operator so that $\det(I + g_{--}^{-1}g_{-+}\tilde{\alpha})$ is well defined and this gives the desired xtension of τ_W to $\mathcal{H}^{\epsilon/2}_-$. In particular if $|\zeta|$ is very small we can take $q_\zeta = 1 - \zeta/z$ and onsider $\tau_W(q_\zeta)$.

7.19) **Proposition.** *Let $\widetilde{\psi}_W(z,f)$ be the function defined in (6.7) then*

7.20)
$$\widetilde{\psi}_W(z,f) = \frac{\tau_W(f\,q_\zeta)}{\tau_W(f)} \cdot \zeta^{-1}.$$

henever $e^{-f}W$ is transversal.

roof. An easy computation shows that

$$\frac{\tau_W(f\,q_\zeta)}{\tau_W(f)} = \tau_{e^{-f}W}(q_\zeta)$$

hus it suffices to see that if W is transversal

$$\widetilde{\psi}_W(\zeta,0) = \tau_W(q_\zeta)\zeta^{-1}$$

et W be the graph of $\alpha : H_- \to H_+$. By definition

$$\widetilde{\psi}_W(\zeta,0) = \zeta^{-1} + \alpha(z^{-1})(\zeta)$$

n the other hand if we set $g = e^{-q_\zeta}$, $a = g_{--}$, $b = g_{-+}$, an easy computation shows that

$$a^{-1}b(h(z)) = \frac{\zeta}{z}h(\zeta), \quad \forall\, h(z) \in \mathcal{H}$$

$a^{-1}b$ has rank 1 and as image the space spanned by z^{-1}. We get that

$$\det(I + a^{-1}b\alpha) = 1 + tr(a^{-1}b\alpha)$$

d

$$\begin{aligned} tr(a^{-1}b\alpha) &= z\left(a^{-1}b\alpha(z^{-1})\right) \\ &= z\frac{\zeta}{z}\alpha(z^{-1})(\zeta) \\ &= \zeta\alpha(z^{-1})(\zeta) \end{aligned}$$

is proves our claim.

Notice that since τ_W is holomorphic, this proves that $\psi_W(z,f)$ is meromorphic in H^W_-.

Using Proposition (7.18) and (7.19) we can thus rewrite the bilinear relation

$$\mathrm{Res}_{z=0}\psi_W(z,t)\psi_{W^\perp}(z,t')dz = 0$$

.21)
$$\mathrm{Res}_{z=0}\frac{\tau_W\left(t_1 - z,\ \ t_2 - \frac{z^2}{2},\ldots\right)\tau_W\left(t_1' + z,\ \ t_2' + \frac{z^2}{2},\ldots\right)}{\tau_W(t)\tau_W(t')}dz = 0$$

where, of course, $f(z)=\sum t_i z^i$. A long but easy computation of residues gives that (7.21 is equivalent to the following set of *Hirota bilinear equations* for τ_W.

$$\boxed{\sum_{j=0}^{\infty}\Delta_j(-2y)\Delta_{j+1}(\widetilde{D}_t)e^{\sum_{k=1}^{\infty}y_k D_{t_k}}\tau(t)\circ\tau(t)=0} \tag{7.22}$$

where

$$\begin{aligned}\sum_{i=0}^{\infty}\Delta_i(x)z^i &= e^{\sum_{j=1}^{\infty}x_j z^j}\\ D_{t_l} &= \frac{\partial}{\partial t_l}\\ \widetilde{D}_{t_l} &= \frac{1}{l}D_{t_l}\\ D_t &= (D_{t_1},D_{t_2},\dots)\\ \widetilde{D}_t &= \left(\widetilde{D}_{t_1},\widetilde{D}_{t_2},\dots\right)\end{aligned}$$

and were, given a differential operator $F(D_{t_l})$,

$$F(D_{t_l})f(t)\circ f(t)=F(D_{y_l})f(t+y)f(t-y)\Big|_{y=0}.$$

Setting $X_i=2i\,t_i$, $D_i=D_{x_i}=\frac{1}{2i}D_{t_i}$, $2y_i=h_i$, we can write (7.22) as

$$\sum_{j=0}^{\infty}\Delta_j(-h)\Delta_{j+1}(2D_i)e^{\sum_{i=1}^{\infty}i h_i D_i}\tau(x)\circ\tau(x)=0. \tag{7.23}$$

We finally show, as we already said, that the space

$$H_-^W=\{f\in H_-,\quad e^{-f}W \text{ is transversal}\}$$

is non empty, for a fixed $W\in Gr(H)$. By the definition of τ this is equivalent to s that $\tau_W(f)$ is not identically 0. To this we fix $W\in Gr(H)$ and an admissible isomorphis $\alpha: H_-\to W$. Set $\tau_W(f)=\tau_l(f)$ where l is the equivalence class of $(\alpha,1)$ in $\mathcal{L}$. S for any sequence $s=(h_1>h_2>\dots)$, such that $h_i=-i$, for $i\gg 0,\alpha^s=j^s\pi\alpha$, whe $j^s: H_-^s\to H_-$ is the isomorphism considered in (6.6). It is then clear that $\alpha^s: H_-\to H$ is admissible so that we can consider

$$P_s(W)=\det(\alpha_s)$$

which is the Plücker coordinate of W relative to the sequence s. Since we have seen th there exists s such that $\pi_-^s: W\to H_-^s$ is an isomorphism it follows that the $P_s(W)$ are n all zero. Using the expression (7.17) we see that

$$\tau_W(f)=\sum_s P_s(W)\tau_{H_-^s}(f)$$

where $\tau_{H_-^s}(f)$ is the τ-function for H_-^s relative to $((j^s)^{-1},1)$. A direct computation [S hows that

$$\tau_{H^{\bullet}_{-}}(t) = \det\left(\Delta_{h_e+\nu-i+j}(t)\right)$$

which is the Schur function relative to the partition $(h_1+1, h_2+2, \dots)$. Since these functions re linearly independent our claim follows.

We finish by giving a very brief sketch of the foundamental link between the τ-function nd the theta-function. This link is established via a construction essentially due to richever. The datum we start with consists in a quintuple (C, p, z, L, φ) where C is a ompact Riemann

surface of genus g, p a point on C, z a local coordinate near p, L a degree $g-1$ line undle on C, and φ a local trivialization of L near p.

Let Δ_ε be a disc of radius ε around p on which z is defined. The Mayer–Vietoris equence

(7.24) $$0 \to \Gamma(C,L) \to \Gamma(C-p,L) \oplus \Gamma(\Delta_\varepsilon, L) \to \Gamma(\Delta_\varepsilon - p, L) \to H^1(C,L) \to 0$$

estricting to Δ_ε identifies $\Gamma(C-p,L)$ and $\Gamma(\Delta_\varepsilon, L)$ with subspaces of $\Gamma(\Delta_\varepsilon - p, L)$. The cal trivialization allows us to substitute L with $\mathcal{O}$, and the local coordinate z identifies $(\Delta_\varepsilon - p, \mathcal{O})$ with a subspace of $H = \lim_{\varepsilon\to 0} \mathcal{O}\left(\dot{\Delta}_\varepsilon\right)$. Taking the limit as ε tends to zero (7.24) we get the exact sequence

$$0 \to \Gamma(C,L) \to \Gamma(C-p,L) \oplus H_+ \to H \to H^1(C,L) \to 0.$$

aking the quotient of the two middle terms by H_+ we set the exact sequence

$$0 \to \Gamma(C,L) \to \Gamma(C-p,L) \xrightarrow{p_-} H_- \to H^1(C,L) \to 0,$$

here p_- is the projection. Since the degree at L is equal to $g-1$, by the Riemann–Roch eorem the dimension of $H^1(C,L)$ is equal to the dimension of $\Gamma(C,L)$. It then follows at p_- is a Fredholm operator of index 0 so that $\Gamma(C-p,L)$ represents a point in the assmannian $Gr_0(H)$. It is also important to notice that the point $\Gamma(C-p,L)$ of $Gr_0(H)$ mpletely determines the quintuple $x = (C, p, z, L, [\varphi])$, where $[\varphi]$ denotes the equivalence ass at φ under the relation $\varphi \sim c\varphi$ for $c \in \mathbf{C}^*$. For instance the curve C and the point p n be reconstructed from $\Gamma(C-p,L)$ as follows

$$C - p = \operatorname{Spec}\left\{f \in H : fW^{\text{alg}} \subset W^{\text{alg}}\right\}$$

here

$$W^{\text{alg}} = \left\{h \in W : h = \sum_{i>k} a_i z^i\right\}.$$

s it was pointed out in [ADKP] the construction of Krichever can be globalized in the llowing way. One can construct an infinite dimensional variety $\widehat{Pic}$, whose points are intuples $x = (C, p, z, L, [\varphi])$ as above, and a holomorphic map

$$\widehat{Pic} \xrightarrow{W} Gr_0(H)$$

fined by: $W(x) = \Gamma(C-p,L) \subset H$. By what we said the map W is injective. Moreover $\widehat{Pic}$ is defined a Poincaré line bundle $\mathcal{O}(\Theta)$ where Θ is the divisor given by

$$\Theta = \{(C, p, z, L, [\varphi]) \quad : \quad \Gamma(C, L) \neq 0\}.$$

Recalling the definition of the determinant bundle $\mathcal{L}$ on the grassmannian $Gr_0(H)$, on easily verifies that $\mathcal{O}(\Theta)$ *is the pull-back via* W *of the dual determinant bundle* $\mathcal{L}^*$. I is then natural to express the τ-function in terms of the theta function. This is done i complete details in [SW]. The final result is the following. Let $x = (C, p, z, L, [\varphi])$, le $\vec{\omega} = (\omega_1, \ldots, \omega_g)$ be a normalized basis at the vector space of holomorphic diffentials, let

$$\int_p^z \vec{\omega} = z\vec{V}_1 + z^2\vec{V}_2 + z^3\vec{V}_3 + \ldots,$$

let $\theta(\vec{\zeta})$, $\vec{\zeta} = (\zeta_1, \ldots, \zeta_g)$, be the Riemann theta function of C then

$$\boxed{\tau_{W(x)} = \theta\left(\sum x_i \vec{V}_i + d_{W(x)}\right) e^{\sum d_{ij} x_i x_j}} \tag{7.25}$$

where

$$\int_p^z \Omega_n = z^{-n} + \sum_0^\infty d_{nj} z^j,$$

Ω_n being the differential of the second kind which has zero a-periods and which is hol morphic expcet for a pole in p with principal part equal to $z^{-n-1}dz$ and where $d_{W(x)}$ the point of the Jacobian of C corresponding to the line bundle L, under a suitable line trasformation of $\mathbf{C}^g$. It is then easy to verify that the system of differential equations (3.3 can be obtained from the Hirota bilinear equations (7.22) via the relation (7.25).

References

[AGR] L. Alvarez-Gaumé, C. Gomez, C. Reina, *Loop groups, Grassmannians and stri theory*, Phys. Lett. B, **190** (1987), 55–62.

[A] E. Arbarello, *Fay's trisecant formula and a characterisation of Jacobian Varieti* Proceedings of Symposia in Pure Mathematics Vol. **46** (1987).

[ACGH] E. Arbarello, M. Cornalba, P.A. Griffiths, J. Harris, *Geometry of algebraic curves*, V I Berlin, Heidelberg, New York: Springer 1985, Vol. **II** (to appear).

[AD1] E. Arbarello and C. De Concini, *On a set of equations characterizing Riemann mat ces*, Ann. of Math. (2) **120** (1984), 119–140.

[AD2] E. Arbarello, C. De Concini, *Another proof of a conjecture of S. P. Novikov on peri and abelian integrals on Riemann surfaces*, Duke Math. J., **54** (1987), 163–178.

[ADKP] E. Arbarello, C. De Concini,, V. Kac, C. Procesi, *Moduli spaces of curves and rep sentation theory*, Comm. Math. Phys, **114** (1988).

[BD] A. Beauville and O. Debarre, *Une rélation entre deux approches du problème de Sch tky*, Preprint.

[BM] A.A. Beilinson, Ju.L. Manin, *The Mumford form and the Polyakov measure in str theory*, comm. Mathy. Phys., **107** (1986), 359–376.

[BMS] A.A. Beilinson, Yu.L. Manin, V.V. Schectman, *Sheaves of the Virasoro and Neveu-Schwarz algebras*, Moscow Univeristy preprint, 1987.

[C] M. Cornalba, *Complex Tori and Jacobians* in Complex Analysis and its applications, Vol. **II** IAEA–SMR 18/24, Vienna 1976.

[DJKM] E. Date, M. Jimbo, M. Kashiwara, T. Miwa, *Transformation groups for soliton equation*, in: *Noalinear Integrabic Systems Classical Theory and Quantum Theory*, World Scientific, Singapore, 1983, 39–119.

[D] R. Donagi, *The Schottky Problem*, in E. Sernesi, *Theory of Moduli*, Lecture Notes Vol. **1337** Springer–Verlag, Berlin–New York 1985.

[Du] B.A. Dubrovin, *Theta functions and non-linear equations*, Russian Math. Surveys, **36**, 2 (1981), 11–92.

[F] J. Fay, *Theta Functions on Riemann Surfaces*, Lecture Notes, Vol. **352**, Springer-Verlag, Berlin–New York, 1973.

[FA] H.M. Farkas, *On Fay's trisecant formula*, J. Analyse Math. **44** (1984), 205–217.

[FR] H. Farkas, H. Rauch, *Period relations of Schottky type on Riemann surfaces*, Ann. Math., **62** (1970), 434–461.

[G] R. C. Gunning, *Some curves in Abelian varieties*, Invent. Math., **66** (1982), 377–389.

[H] J. Harer, *The second homology group of the mappings class group of an orientable surface*, Invent. Math., **72** (1983), 221–239.

[K] I.M. Krichever, *Methods of Algebraic Geometry in the theory of nonlinear equations*, Russian Math. Surveys, **32** (1977), 185–213.

[KP] V.G. Kac, D.H. Peterson, *Spin and wedge representations of infinite dimensional Lie algebras and groups*, Proc. Nat. Acad. Sci. U.S.A., **78** (1981), 3308–3312.

[KNTY] N. Kawamoto, Y. Namikawa, A. Tsuchiva, Y. Yamada, *Geometric realization of conformal field theory on Riemann surfaces*, Comm. Math. Phys. (in press).

[I] J. Igusa, *Theta Functions*, Grundlehren der Math. Wiss., **194**, Springer, 1972.

[M] Yu. Manin, *Quantum string theory and algebraic curvers*, Berkeley J.C.M. Talk, 1986.

[M1] D. Mumford, *Curves and their Jacobians*, Ann. Arbor, University of Michigan Press, 1975.

[M2] D. Mumford, *An enumerative geometry of the moduli space*, in *Arithmetic and Geometry*, papers dedicated to I.R. Shafarevich, Birkhaser, Boston, 1983.

[M3] D. Mumford, *An Algebro-geometric construction of commuting operators and of solutions to the Toda lattice equation, Korteweg-de Vries equation and related non-linear equations*, (Proc. Internat. Sympos. Alg. Geometry, Kyoto, 1977), Kinokuniy Book Store, Tokyo, 1978, 115–153.

[M4] D. Mumford, *Tata lectures on theta. II*, Birkhaüser, Boston, 1983.

[MF] D. Mumford, J. Fogarty, *Geometric invariant theory*, Ergebnisse der Math., **34**, Springer.

[Mu] M. Mulase, *Cohomological structure in soliton equations and Jacobian varieties*, J. Differential Geom., **19** (1984), 403–430.

[N] S.P. Novikov, *The periodic problem for the Korteweg-de Vries equation*, Functional Anal. Appl. **8** (1974), 236–246.

[OS] F. Oort, J. Steenbrink, *The local Torelli problem for algebraic curves*, Journées de Géometrie Algébrique d'Angers, juillet 1979, 157–204, Sijthoff and Noordhoff, Alphen aan den Rijn, 1980.

[PS] A. Pressley, G. Segal, *Loop Groups*, Oxford University press, 1986.

[R] A.K. Raina, *Fay's trisecant identity and conformal field theory*, preprint TIFR/TH/88-37.

[SW] G. Segal, G. Wilson, *Loop groups and equations of KdV type*, Publ. Math., K.E.S. **61**

[Sh] T. Shiota, *Characterization of Jacobian varieties in terms of soliton equations*, Invent Math., **83** (1986), 333–382.

[vG2] B. Van Geemen, *The Schottky problem and moduli spaces of Kummer varieties*, U. o Urecht thesis, 1985.

[W] W. Wirtinger, *Untersuchungen über Thetafunctionen*, Teubner, Berlin, 1985.

[W1] G. Welters, *A criterion for Jacobi varieties*, Ann. Math. **120** (1984), 497–504.

[W2] G. Welters, *On flexes of the Kummer variety*, Nederl. Akad. Wetensch. Proc. Ser A 86 **45** (1983), 501–520.

GEOMETRY OF STANDARD CONSTRAINTS AND ANOMALOUS SUPERSYMMETRIC GAUGE THEORIES†

Ugo Bruzzo

Dipartimento di Matematica, Università di Genova,

Via L. B. Alberti 4, I-16132 Genova, Italy

Giovanni Landi ‡

Istituto Nazionale di Fisica Nucleare, Sezione di Napoli,

Mostra d'Oltremare Pad. 19, I-80125 Napoli, Italy

ABSTRACT. Supermanifold theory is used to give a geometric interpretation of the standard constraints which are imposed in superspace formulations of supersymmetric gauge theories. The results obtained are exploited, together with a few facts from supermanifold cohomology, to provide a simple proof of Weil triviality in anomalous supersymmetric gauge theories.

INTRODUCTION

When formulating supersymmetric theories on superspace, one usually imposes constraints to ıminate redundant degrees of freedom. In pure supersymmetric Yang-Mills theories one imple- ents the *standard constraints*, namely one requires the vanishing of the supercurvature compo- nts when both indices are spinor-like, $F_{\alpha\beta} = 0$.[1)]

Standard constraints have been used in Ref. 2 to prove Weil triviality in supersymmetric uge theories. Weil triviality is in a sense a sufficient condition in order that the consistent gauge Lorentz anomalies of these theories can be expressed as polynomial functionals of the gauge tential and of its field strength.

In this paper we use techniques of superfibre bundle theory to give a geometric interpretation the standard constraints. We then use supermanifold cohomology techniques to provide a very nple proof of Weil triviality.

This paper is organized as follows. In Section 2 we review some basic notions of supermanifold eory, including some cohomology. Then, given a principal superfibre bundle, we use a connection it to lift to the total space some suitable subbundles of the graded tangent bundle to the base permanifold. In the physical situations, the standard constraints will be related to geometric operties of the lifted subbundles. In section 3, after a quick review of the anomaly problem, describe our proof of Weil triviality. Finally, we comment on the possibility of integrating omalies over space-time.

† Talk given by the second author.

‡ Also at SISSA, Strada Costiera 11, I-34014 Trieste, Italy.

2. GEOMETRIC PRELIMINARIES

2.1. Supermanifolds

In this paragraph we give a few basic notions of the so-called "geometric" theory of super manifolds, initiated by De Witt and Rogers.[3,4] For further developments of the theory, whic eliminated some drawbacks of previous approaches, the reader may consult Refs. 5,6. We star with a real exterior algebra B_L with L generators, endowed with its natural structure of $\mathbb{Z}_2$-grade commutative algebra, † so that $B_L = (B_L)_0 \oplus (B_L)_1$. An (m,n) dimensional supermanifold M i a topological space locally modelled over the 'vector superspace'

$$B_L^{m,n} = (B_L)_0^m \times (B_L)_1^n$$

by using an atlas whose transition functions are 'supersmooth' functions. A supersmooth functic $f : U \subset B_L^{m,n} \to B_L$ has the usual form (superfield expansion)

$$f(x^1 \dots x^m, \theta^1 \dots \theta^n) = f_0(x) + \sum_{\alpha=1}^{n} f_\alpha(x)\,\theta^\alpha + \dots + f_{1\dots n}(x)\,\theta^1 \dots \theta^n$$

where the x's are the even (Grassmann) coordinates, the θ's are the odd ones, and the dependence the coefficient functions $f_{\dots}(x)$ on the even variables is fixed by their behaviour for real argument

Let $\mathcal{G}$ be the sheaf of germs of supersmooth B_L-valued functions on M, and $Der^*\mathcal{G}$ the du of the sheaf $Der\mathcal{G}$ of graded derivations of $\mathcal{G}$. We can define the sheaf Ω^k of supersmooth k-forn (or k-superforms) on M to be

$$\Omega^k = \wedge_{\mathcal{G}}\, Der^*\mathcal{G} \qquad k \geq 1.$$

The exterior differential $d : \Omega^k \to \Omega^{k+1}$ is defined in the usual way. The *supersmooth de Rha sheaf complex*

$$\mathcal{G} \xrightarrow{d} \Omega^1 \xrightarrow{d} \Omega^2 \to \dots ,$$

is exact, that is, there is a Poincaré Lemma for superforms.[7]

If $\Gamma\Omega^k$ is the space of global sections of Ω^k, i.e. the graded B_L-module of global k-superforr on M, the cohomology $H^*[\Gamma\Omega^*]$ is called the *supersmooth de Rham cohomology* of M, denoted $H^*_{SDR}(M)$.[8,9] Since in general the sheaves $\mathcal{G}$ and Ω^k have non-trivial Čech cohomology in degr higher than zero,[9] there is no analogue of de Rham's theorem

$$H^k_{SDR}(M) \simeq H^k(M, B_L), \qquad k \geq 0, \qquad ($$

where $H(M, B_L)$ is the Čech cohomology of the constant sheaf B_L on M.

However, there is a class of physically relevant supermanifolds for which the isomorphism (holds true, i.e. the so-called De Witt supermanifolds.[3] An (m,n) dimensional supermanifold is

† In the following by "graded" we shall always mean "$\mathbb{Z}_2$-graded."

Vitt if it is a locally trivial bundle over an m-dimensional ordinary smooth manifold M_0, called the *ody* of M, with fibre the $\mathbb{R}$-vector space $N_L^{m,n} = (N_L)_0^m \times (N_L)_1^m$, N_L being the ideal of nilpotent lements of B_L. If M is De Witt, the sheaves $\mathcal{G}$ and Ω^k have trivial Čech cohomology,[10] and this 1 turn implies that the isomorphism (1) holds. Since a De Witt supermanifold is contractible to s body, de Rham's theorem for M_0 implies

$$\begin{aligned} H^k_{SDR}(M) &\equiv H^k_{DR}(M_0) \otimes B_L \quad \text{for} \quad 0 \le k \le m, \\ H^k_{SDR}(M) &= 0 \quad \text{for} \quad k > m, \end{aligned} \tag{2}$$

rovided that M is a De Witt supermanifold with even dimension m.

.2. Geometry of Standard Constraints

The graded tangent bundle TM to an (m,n) dimensional supermanifold M is a supervector undle of rank (m,n). We assume that TM has a direct sum splitting $TM = T'M \oplus T''M$, with rojections $p_0 : TM \to T'M$, $p_1 : TM \to T''M$, and define a morphism of super vector bundles $: TM \to TM$ by setting $J \doteq p_0 - p_1$. One has $J^2 = 1$, i.e. J is a graded involution. A local ection X of TM is of type (1,0) (resp. (0,1)) if $J(X) = X$ (resp. $J(X) = -X$). The *torsion nsor* (or Nijenhuis tensor) of J is the even (1,2) tensor field defined by

$$N(X,Y) = [X,Y] + [JX,JY] - J[JX,Y] - J[X,JY]$$

here X, Y are graded derivations and the commutators are graded.

Denoting by $T^*M = T^{*\prime}M \oplus T^{*\prime\prime}M$ the dual splitting of the graded cotangent bundle, we efine the supervector bundles of superforms of type (p,q):

$$\Omega^{p,q} = \left(\wedge^p_{\mathcal{G}} T^{*\prime}M\right) \otimes_{\mathcal{G}} \left(\wedge^q_{\mathcal{G}} T^{*\prime\prime}M\right).$$

hen there is a decomposition

$$\Omega^k = \bigoplus_{p+q=k} \Omega^{p,q}$$

th projections $\pi^{p,q} : \Omega^{p+q} \to \Omega^{p,q}$. A section η of $\Omega^{p,q}$, i.e. a superform of type (p,q) over an en set U in M, is locally expressed as

$$\eta = \sum \eta_{i_1 \ldots i_p, \alpha_1 \ldots \alpha_q}\, \omega^{i_1} \wedge \ldots \wedge \omega^{i_p} \wedge \omega^{\alpha_1} \wedge \ldots \wedge \omega^{\alpha_q} \tag{3}$$

th the ω^i's of type (1,0) and the ω^α's of type (0,1). Considering the exterior differential as a eaf map $d : \Omega^{p,q} \to \Omega^{p+q+1}$, we define

$$d_0 = \pi^{p+1,q} \circ d, \quad d_1 = \pi^{p,q+1} \circ d, \quad T = \pi^{p-1,q+2} \circ d, \quad \tau = \pi^{p+2,q-1}.$$

om the representation (3) one gets

$$d = d_0 + d_1 + T + \tau, \tag{4}$$

i.e.

$$d\Omega^{p,q} \subset \Omega^{p+1,q} \oplus \Omega^{p,q+1} \oplus \Omega^{p-1,q+2} \oplus \Omega^{p+2,q-1} \, .$$

The appearance of T (resp. τ) in the decomposition (4) of d is equivalent to the fact that $T'M$ (resp. $T''M$) is not involutive. One has indeed the following results.[11]

Proposition 1. *The involutivity of $T'M$ is equivalent to each of the following conditions:*

(i) $d\Omega^{0,1} \subset \Omega^{1,1} \oplus \Omega^{0,2}$;

(ii) $d\Omega^{p,q} \subset \Omega^{p+1,q} \oplus \Omega^{p,q+1} \oplus \Omega^{p-1,q+2}$;

(iii) $d = d_0 + d_1 + T$.

Analogously, the simultaneous involutivity of $T'M$ and $T''M$ is equivalent to each of the following conditions:

(i) $d\Omega^{0,1} \subset \Omega^{1,1} \oplus \Omega^{0,2}$, $\qquad d\Omega^{1,0} \subset \Omega^{2,0} \oplus \Omega^{1,1}$;

(ii) $d\Omega^{p,q} \subset \Omega^{p+1,q} \oplus \Omega^{p,q+1}$;

(iii) $d = d_0 + d_1$;

(iv) $N = 0$.

Since in the next section we shall be concerned with supersymmetric gauge theories, we consider principal superfibre bundle $\pi : Q \to M$ with structure (super)group G.[12,13,14] Assuming that th graded tangent bundle to M has a direct sum splitting, we shall use a connection ∇ on Q to li that splitting to TQ. If F is the curvature form of the connection, we shall relate the conditi $F^{0,2} = 0$ (which in physical applications corresponds to the standard constraints) to geomet properties of the subbundles of TQ so obtained. We regard a connection ∇ on Q as a G-invaria splitting of the exact sequence of super vector bundles over Q:[12,15,16]

$$0 \to T^V Q \to TQ \to \pi^* TM \to 0, \qquad ($$

$T^V Q$ being the *vertical graded tangent bundle* whose sections are vertical graded derivations Q. For each $u \in Q$ the connection yields an isomorphism

$$T_u Q \simeq T_{\pi(u)} M \oplus T_u^V Q.$$

The splitting $TQ = T'Q \oplus T''Q$ is obtained by identifying by means of the isomorphism (6) T' with $T'M$ plus the vertical part of TQ, while $T''Q$ is identified with $T''M$, so that

$$T'_u Q \simeq T'_{\pi(u)} M \oplus T^V_u Q, \qquad T'' Q \simeq T''_{\pi(u)} M.$$

The subbundles $T'Q$ and $T''Q$ enjoy the following properties.

Proposition 2. *$T'Q$ is involutive if $T'M$ is. Moreover, provided that $T'M$ and $T''M$ are bo involutive, the following conditions are equivalent:*

(i) $T''Q$ is involutive;

(ii) for any pair of graded derivations X, Y on Q of type (0,1), $[X,Y]$ is horizontal;
(iii) $F^{0,2}=0$, where F is the curvature form of the connection ∇.

If $T''M$ is not involutive, conditions (ii) and (iii) 4 are still equivalent.

. WEIL TRIVIALITY

In this section we apply the techniques of the previous section to provide a simple proof of 'eil triviality in supersymmetric gauge theories. As it is well known by now,[2] Weil triviality is a ıfficient condition for generalizing to supersymmetric theories the differential geometric methods ɛe the transgression formula, which are exploited for the analysis of anomalies in ordinary Yang-ills theories, in particular to obtain polynomial expressions of the anomalies. Weil triviality was iginally proved in Ref. 2 by means of different methods, which for a high space-time dimension involve a cumbersome and intriguing analysis of the representations of the group $SO(m-1,1)$.

1. The anomaly problem

We give a brief outline of the anomaly problem in supersymmetric gauge theory. To be definite consider the case of supersymmetric chiral anomalies for a super Yang-Mills field coupled with external simple supergravity field. Let Q be a principal superfibre bundle over an (m,n) nensional supermanifold M with structure supergroup G, and suppose on Q there is a connection with connection form ω and curvature form F. For simplicity we assume that the bundle Q trivial, but everything can be easily generalized to the case of a nontrivial bundle by using a ckground connection.[17]

In order to write the BRST transformations we need the Faddeev-Popov ghost c of the group gauge transformations, which locally is a mapping from M into Lie(G), and the ghost $\xi=\xi^A\partial_A$ the superdiffeomorphism group; here A is a collective index $A=(a,\alpha)$, $a=1\ldots m$, $\alpha=1\ldots n$. en the BRST transformations can be locally written as

$$\begin{aligned} &\delta_G\omega=-D_\omega\, c, && \delta_G c=-c^2, && \delta_G\xi=0 \\ &\delta_D\omega=\mathcal{L}_\xi\omega, && \delta_D c=\mathcal{L}_\xi c, && \delta_D\xi=\xi^B\partial_B\xi^A. \end{aligned} \tag{6}$$

The BRST operators δ_G, δ_D, $\delta=\delta_G+\delta_D$ are nilpotent, thus giving rise to cohomology ories (*BRST cohomologies*). The degrees in the differential complexes underlying the BRST ıomologies are called *ghost numbers*.

An anomaly $\mathcal{A}$ is a non trivial δ-cohomology class modulo d, with ghost number one, namely

$$\delta\mathcal{A}=d\mathcal{B} \tag{7}$$

for some $\mathcal{B}$, with $\mathcal{A} \neq \delta\mathcal{A}' + d\mathcal{B}'$. Only BRST cohomology classes modulo d are considered for o is interested in the space-time integral of the anomaly; since space-time is assumed to be compa without boundary, exact forms do not contribute.

If one writes $\mathcal{A} = \mathcal{A}_D + \mathcal{A}_G$, with $\mathcal{A}_D$ and $\mathcal{A}_G$ linear in the corresponding ghost fields, th Eq. (7) implies that the forms $\delta_D \mathcal{A}_D$, $\delta_D \mathcal{A}_G + \delta_G \mathcal{A}_D$, $\delta_G \mathcal{A}_G$ are all d-exact. This property is t consistency condition for the anomalies $\mathcal{A}_G$ and $\mathcal{A}_D$. One also requires that the anomalies are loc expressions of the connection ω and of its curvature F.

Now, let us take an invariant polynomial P of order $k = \frac{1}{2}m + 1$ on the Lie (super)algebra G. The BRST equations (6) entail

$$P(F^k) = (d + \delta_G)\left\{k \int_0^1 dt\ P(\omega', \mathcal{F}_t^{k-1})\right\} \doteq (d + \delta_G)S, \qquad ($$

where $\omega' = \omega + c$, $\omega'_t = t\omega'$ and $\mathcal{F}_t = (d + \delta_G)\omega'_t + \frac{1}{2}[\omega'_t, \omega'_t]$. In Ref. 2 it has been shown th anomalies $\mathcal{A}_G$, $\mathcal{A}_D$ fulfilling the consistency conditions can be obtained from the descent equatio determined by P provided that

$$P(F^k) = dX, \qquad \delta_G X = 0. \qquad (9a,$$

Under these conditions, Eq. (8) can be written as

$$(d + \delta_G)\hat{S} = 0, \qquad \hat{S} = S - X. \qquad (1$$

By expanding Eq. (10) according to the ghost number, one gets a string of equations

$$d\hat{S}^0_{2k-1} = 0, \quad dS^1_{2k-2} + \delta_G \hat{S}^0_{2k-1} = 0, \quad dS^2_{2k-3} + \delta_G S^1_{2k-2} = 0, \quad \cdots$$

where S^q_h is a h-superform with ghost number q. If one now defines

$$\mathcal{A}_G = S^1_{2k-2}, \quad \mathcal{A}_D = -i_\xi \hat{S}^0_{2k-1},$$

one can prove that

$$\delta_G \mathcal{A}_G = -dS^2_{2k-3},$$

$$\delta_D \mathcal{A}_G + \delta_G \mathcal{A}_D = di_\xi S^1_{2k-2},$$

$$\delta_D \mathcal{A}_D = 2di_\xi i_\xi \hat{S}^0_{2k-1},$$

so that the $(2k-2)$-superforms $\mathcal{A}_G$, $\mathcal{A}_D$ solve the problem (7).

If condition the conditions (9) were not satisfied it would have been impossible to fin partner $\mathcal{A}_D$ of $\mathcal{A}_G$ such that the consistency conditions are satisfied.[2)] The property (9), toget with the requirement that X is local in the components of the potential field (connection) and the field strength (curvature), is called *Weil triviality*.

3.2. Exactness of $P(F^k)$

From a physical point of view, it does not seem to be restrictive to assume that the supermanifold M is De Witt; indeed, so far no physical application of non-De Witt supermanifolds is known. Since the $2k$-superform $P(F^k)$ is closed,[2,12] from (2) it follows that it is also exact for all $k > \frac{m}{2}$. We stress that in order to prove this result no constraints on the curvature are needed.

3.3. Locality of the form X

Using the techniques described in Section 2 we can show that the form X appearing in Eq. (9), which is determined only up to closed forms, can be chosen so as to be local in the components of the connection form. The splitting of the tangent space to the base supermanifold of the principal fibration we are considering is introduced in terms of the superfibre bundle $Lor(M)$, which is a subbundle of the superbundle $L(M)$ of frames over M. The bundle $Lor(M)$ is a natural object to consider in the case of a supersymmetric gauge theory coupled with supergravity and invariant under the local action of the Lorentz group. Assuming that M is (m,n) dimensional, the structure group of $Lor(M)$ is (the Grassmannification of) $Spin(m-1,1)$, the covering group of $SO(m-1,1)$. The structure group acts on the frames by means of matrices of the form

$$\begin{pmatrix} \alpha(S) & 0 \\ 0 & \beta(S) \end{pmatrix}, \qquad S \in Spin(m-1,1). \tag{11}$$

where the map $\alpha : Spin(m-1,1) \to SO(m-1,1)$ is the covering homomorphism, and the map $\beta : Spin(m-1,1) \to GL(n)$ is a suitable spin representation.

Let $\sigma = \{D_i, D_\alpha,\, i = 1\dots m,\ \alpha = 1\dots n\}$ be a section of $Lor(M)$ over an open set $U \subset M$, and let $\Gamma_i{}^k$, $i,k = 1\dots m$ be the $so(m-1,1)$-valued connection form of a connection on $Lor(M)$. We assume the usual constraints on the torsion:[1)]

$$\sigma^* T^i = -(C\gamma^i)_{\alpha\beta}\,\omega^\alpha\,\omega^\beta, \qquad \sigma^* T^\alpha = 0$$

where $T^A = \{T^i, T^\alpha\}$ is the torsion form of the connection, the ω's are the co-frames dual to the D's, and C is the charge conjugation matrix. Denoting by $\Gamma_{Ai}^{\ k}$ the local components of the connection form, $\sigma^*\Gamma_i{}^k = \omega^A\,\Gamma_{Ai}^{\ k}$, $A = 1\dots m+n$, the graded commutators of the frame fields have the following expressions:

$$\begin{aligned}
[D_i, D_k]_- &= (\Gamma_{ki}^{\ j} - \Gamma_{ik}^{\ j}) D_j, \\
[D_i, D_\alpha]_- &= \Gamma_{\alpha_i}^{\ j}\, D_j + (\Sigma^j{}_k)^\beta{}_\alpha\, \Gamma_{ij}^{\ k}\, D_\beta \\
[D_\alpha, D_\beta]_+ &= (C\gamma^i)_{\alpha\beta}\, D_i + \tfrac{1}{2}\left((\Sigma^i{}_k)^\lambda{}_\alpha \Gamma_{\beta i}^{\ k} + (\Sigma^i{}_k)^\lambda{}_\beta \Gamma_{\alpha i}^{\ k}\right) D_\lambda
\end{aligned} \tag{12}$$

where $\Sigma_{ik} = \frac{1}{4}[\gamma_i, \gamma_k]$. If the gauge theory is not coupled to supergravity, only the term $(C\gamma^i)_{\alpha\beta} D_i$ survives in Eqs. (12) and the Lie superalgebra of global supersymmetry is recovered.

In any case, the D_i's generate over the supersmooth functions on U an involutive rank $(m, 0$ subbundle of TU, whilst the D_α's generate a non-involutive rank $(0, n)$ subbundle. Since th structure group of $Lor(M)$ acts in the block form (11), these local subbundles do not interming and glue together to yield a global splitting $TM = T'M \oplus T''M$.

As described in Section 2.2, that splitting can be lifted to the total space of the principa superfibre bundle $\pi : Q \to M$, so that $TQ = T'Q \oplus T''Q$. Locally the spaces of sections of $T'Q$ are spanned by the vector fields $\{D_i^\star, \xi_A\}$, where the $D_i^\star$ are the horizontal lifts of the D_i and th ξ_A are a basis of fundamental vector fields associated with the action of G; the spaces of section of $T''Q$ are generated by the horizontal lifts $D_\alpha^\star$. $T'Q$ is involutive, while obviously $T''Q$ is no The condition $F^{0,2} = 0$ involved in Proposition 2 is locally expressed by $F_{\alpha\beta} = 0$, i.e. it yields th so-called standard constraints. According to Proposition 1, the involutivity of $T'Q$ is equivalent t a decomposition $d = d_0 + d_1 + T$ of the exterior differential on Q. The operator T coincides wit the operator denoted by the same symbol in Ref. 2; it can be locally expressed as

$$T\eta = (C\gamma^i)_{\alpha\beta}\, \omega^{\star\alpha}\, \omega^{\star\beta}\, i_{D_i^\star}\eta, \tag{13}$$

where the $\omega^{\star i}$, $\omega^{\star\alpha}$ are 1-forms dual to $\{D_i^\star, D_\alpha^\star, \xi_A\}$. The condition $d^2 = 0$ splits into

$$\begin{aligned} T^2 &= 0 \\ d_1 T + T d_1 &= 0 \\ d_0 T + T d_0 + d_1^2 &= 0 \\ d_0 d_1 + d_1 d_0 &= 0 \\ d_0^2 &= 0 \end{aligned} \tag{14}$$

We can now prove the locality of the form X appearing in Eq. (9). To be definite, we assur that m (which may be identified with the space-time dimension) is four, but the same results ho for $m = 2, 6, 10$; in particular for $m = 2$ the standard constraints are not necessary.[18,19]

In the case $m = 4$ one takes $k = 3$. Due to the standard constraints, $P(F^3)$ has only the (4, and (3,3) terms:

$$P(F^3) = P^{4,2} + P^{3,3}.$$

One easily checks that any X satisfying $P(F^3) = dX$ has only terms of type (4,1), (3,2), and (2, Now, it is possible to find a 4-superform ς such that $\tilde{X} = X + d\varsigma$ is of type (4,1). To prove th we must show that the equations

$$\pi^{3,2}(X + d\varsigma) = 0, \qquad \pi^{2,3}(X + d\varsigma) = 0, \qquad \pi^{1,4}(d\varsigma) = \pi^{0,5}(d\varsigma) = 0$$

admit solutions. Simple local calculations show that this is indeed the case.[19]

Therefore we may assume that X is of type (4,1). Then condition $P(F^3) = dX$ splits into

$$P^{4,2} = d_1 X, \qquad P^{3,3} = TX. \tag{15a}$$

As a final step to prove our result, we need to introduce the operator $S : \Omega^{p,q} \to \Omega^{p+1,q-2}$ locally defined by

$$S\eta = \frac{1}{2n}(\gamma_i C^{-1})^{\alpha\beta}\, \omega^i\, i_{D_\alpha} i_{D_\beta} \eta.$$

A simple calculation shows that, if η is of type (p,q), then

$$(TS+ST)\eta = p\eta + \frac{1}{2n}(C\gamma^i)_{\alpha\beta}\, (\gamma_i C^{-1})^{\mu\nu}\, \omega^\alpha\, \omega^\beta\, i_{D_\mu} i_{D_\nu}\, \eta. \tag{16}$$

Applying the operator S to (15b) one obtains, as a consequence of Eq. (16),

$$X = \tfrac{1}{4} S P^{3,3}$$

so that X is local, since $P^{3,3} \equiv \pi^{3,3}(P(F^3))$ is a polynomial in the components of the connection and their derivatives.

The gauge invariance of the form X is equivalent to the condition $\mathcal{L}_Z X = 0$, with Z is a vertical G-invariant graded vector field on P.[15,16] Since both $P(F^3) = dX$ and X are horizontal, one has

$$\mathcal{L}_Z X = Z \rfloor dX + d\,(Z \rfloor X) = 0.$$

This completes the proof of Weil triviality.

.4 Integrating Anomalies over Spacetime

In supersymmetric gauge theory the anomalies $\mathcal{A}_G$ and $\mathcal{A}_D$ determined in section 3.1 are usually integrated over space-time by means of a formal procedure, cf. Ref 2. Actually, it is possible to give a precise mathematical meaning to those integrals, provided that the integrands are supersymmetric quantities. Let M be an (m,n) dimensional De Witt supermanifold with body M_0 a compact orientable manifold without boundary. Given an m-superform η on M, it can be integrated over M_0 by pulling it back by means of a global section $\sigma : M_0 \to M$ of the smooth bundle $M \to M_0$, which always exists since the fibre of the bundle is diffeomorphic to a vector space. Thus the integral

$$\int_{M_0} \sigma^* \eta \tag{17}$$

is well defined, but it depends also on the section σ. In Ref. 20 it was shown that an integral like (17) does not depend upon σ provided that its integrand is invariant under local supersymmetry transformations up to an exact form. Thus the space-time integrals of the anomalies are well defined whenever one can choose a supersymmetric representative of the anomaly cocycles.

Acknowledgments. The research on which this work is based was done partly in collaboration with C. Bartocci, which we would like to thank, and was partially supported by 'Gruppo Nazionale per la Fisica Matematica' of the Italian Research Council, by 'Istituto Nazionale di Fisica Nucleare', Italy, and by the Italian Ministry for Public Education through the research project 'Geometria e Fisica'.

REFERENCES

1. Witten, E. Nucl. Phys. **B266**, 245 (1986); and references therein.
2. Bonora, L., Pasti, P. and Tonin, M., Nucl. Phys. **B286**, 150 (1987); and in 'Field an Geometry,' A. Jadczyk ed. (Singapore, World Scientific 1987).
3. De Witt, B., 'Supermanifolds' (London, Cambridge Univ. Press 1984).
4. Rogers, A., J. Math. Phys. **21**, 1352 (1980); Commun. Math. Phys. **105**, 375 (1986).
5. Bartocci, C. and Bruzzo, U., J. Geom. Phys. **4**, 391 (1987).
   Bartocci, C., Bruzzo, U. and Hernández Ruipérez, D., "A remark on a new category of supe manifolds," Preprint, Dip. di Matematica Univ. di Genova.
6. Rothstein, M., Trans. Amer. Math. Soc. **297**, 159 (1986).
7. Bruzzo, U., in "Differential Geometric Methods in Theoretical Physics," K. Bleuler and M Werner eds. (Kluwer, to appear).
8. Rabin, J. M., Commun. Math. Phys. **108**, 375 (1987).
9. Bartocci, C. and Bruzzo, U., J. Math. Phys. **28**, 2363 (1987).
10. Bartocci, C. and Bruzzo, U., J. Math. Phys. **29**, 1789 (1988).
11. Bartocci, C., Bruzzo, U. and Landi, G., "Geometry of Standard Constraints and Weil Triviali in Supersymmetric Gauge Theories," Preprint 65/1988, Dip. di Matematica Univ. di Genov
12. Bartocci, C., Bruzzo, U. and Landi, G., "Chern-Simons Forms on Principal Super Fib Bundles," Preprint SISSA 109/87/FM, Trieste 1987.
13. Rogers, A., J. Math. Phys. **22**, 939 (1981).
14. Rittenberg, V., Scheunert, M., J. Math. Phys. **19** 713 (1978).
15. Atiyah, M. F. and Bott, R., Phil. Trans. R. Soc. London **A308**, 523 (1982).
16. López Almorox, A., in "Differential Geometric Methods in Mathematical Physics," P. García and A. Pérez-Rendón eds., Lect. Notes Math. **1251** (Berlin, Springer-Verlag 1987).
17. Mañes, J., Stora, R. and Zumino, B., Commun. Math. Phys. **102**, 157 (1985).
18. Buckingham, S., "Weil Triviality and Anomalies in Two Dimensional Supergravity," Kin College Preprint, London, May 1987.
19. Bruzzo, U. and Landi, G., "A Simple Proof of Weil Triviality in Supersymmetric Gau Theories," Preprint 64/1988 Dip. di Matematica Univ. di Genova.
20. Bruzzo, U., and Cianci, R., Commun. Math. Phys. **95**, 393 (1984).

HAMILTONIAN FORMULATION OF STRING THEORY AND MULTILOOP AMPLITUDES IN THE OPERATOR CONTEXT

Adrian R. Lugo
Jorge G. Russo
International School for Advanced Studies
Strada Costiera 11, Trieste, Italy

Abstract:

The operator formalism for string theory at arbitrary genus is presented in great detail. A Hamiltonian operator is provided. This dictates the time evolution of any operator of the theory and, in particular, allows us to derive the equations of motion of the fundamental fields. Scattering amplitudes are defined as correlation functions of suitable vertex operators. The formalism lets one in general compute any correlation function. We compute correlation functions involving the matter field and reobtain standard results.

1.Introduction

The operator formalism in string theories is well known at genus zero and was extensively studied in the literature [1]. At higher genus, however, there is no systematic operator treatment and computations are more frequently made by using path integral techniques [2]. Recently there has been an increasing interest in developing operator methods on higher genus Riemann surfaces [3,4]. Though interesting, these approaches are not as simple and natural as one would wish. The aim of this seminare is to show how one can introduce a very natural and elegant operator formalism which closely follows the lines of the genus zero case.

In ref.[5] Krichever and Novikov (KN) showed that it is possible to provide explicitely bases for the space of meromorphic tensor fields of weight λ, holomorphic outside two distinguished points P_+ and P_-. These kind of bases can be constructed by simply using the Riemann-Roch theorem. Of particular interest is the case $\lambda=-1$, namely, vector fields. They can be used to generate as well diffeomorphisms as Teichmüller deformations of the Riemann surface. They satisfy an algebra (called KN algebra) which is the generalization of the Virasoro algebra to higher genus. In ref.[4] they used these bases to introduce an operator formalism for string theory. The explicit construction of these bases was made in ref.[6], where a study of b-c systems is performed in this context.

This talk is based on ref.[7]. We found convenient to introduce some changes in the notation. Beside, there are other important modifications due to errors of ref.[7] which have been corrected.

We are grateful to Prof.M. Francaviglia for allowing us to participate in this 1988 CIME course.

2. Construction of an operator formalism on genus g Riemann surfaces.

Let us recall some elementary facts of genus zero. When a string propagates in space-time it sweeps out a world-sheet which is topologically a cylinder (Fig.1a). This is conventionally parametrized by an angular coordinate σ and a time evolution parameter t. By going to euclidean time, τ ($=it$), this cylinder can be mapped to the complex plane without the points $z=0$ and $z=\infty$ by simply defining $z=\exp(\tau+i\sigma)$ as the coordinate of the complex plane (Fig.1b). This is conformal to a sphere without two points (Fig.1c). The inverse map can be defined by

$$\tau(z,\bar{z}) = \mathrm{Re}\int_1^z dz/z \tag{2.1a}$$

$$\sigma(z,\bar{z}) = \mathrm{Im}\int_1^z dz/z \quad , \ \sigma \simeq \sigma + 2\pi n, \ n \in N \tag{2.1b}$$

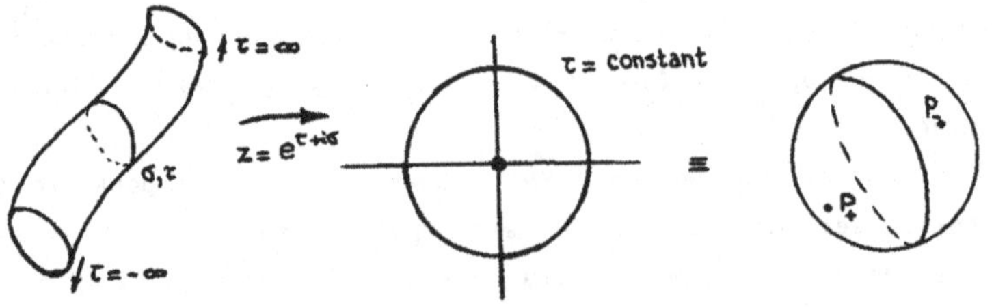

Fig.1a: Cylinder. Fig.1b: Complex plane without $z=0$, ∞. Fig.1c: Sphere without two points.

We see that the level curves of equal τ (which represent the string) are concentric circles around $z=0$ (Fig.1b).

At higher genus we have something similar: The string propagates but this time it splits and then joins giving rise to "holes" which topologically characterize the two dimensional surface (Fig.2a). Again, by going to euclidean time, this can be conformally mapped to a Riemann surface Σ without two points P_+ and P_- (Fig.2b). Likewise one defines

$$\tau(P) = \mathrm{Re}\int_{P_o}^{P} dk \quad ; \; P,\; P_o \in \Sigma \tag{2.2}$$

where dk is a differential of the third kind with simple poles at P_+ and P_- with residues +1 and -1 respectively. This defines dk up to addition of holomorphic differentials. If we require τ to be single-valued, that is

$$\mathrm{Re}\oint_{\gamma_i} dk = 0 \tag{2.3}$$

with γ_i any homology cycle, then dk is fixed unambiguously:

$$dk(P) = d[\log(E(P,P_+)/E(P,P_-)] - i2\pi \sum_{i,j=1}^{g} \mathrm{Im}(\int_{P_-}^{P_+} \eta_i)(\mathrm{Im}\Omega)^{-1}_{ij}\, \eta_j$$

Thus

$$\tau(P) = \mathrm{Re}(\, \log(\frac{E(P,P+)E(Po,P-)}{E(P,P_-)E(P_o,P_+)}) - i2\pi \sum_{i,j=1}^{g} \mathrm{Im}(\int_{P_-}^{P_+}\eta_i)(\mathrm{Im}\Omega)^{-1}_{ij} \int_{P_o}^{P} \eta_j \,)$$

where $\{\eta_i\}$ is the basis of holomorphic differentials normalized around the standard basis of homology (α_i,β_i) according to

$$\oint_{\alpha_i} \eta_i = \delta_{ij} \quad , \quad \oint_{\beta_i} \eta_i = \Omega_{ij}$$

By analogy with the g=0 case one could define

$$\sigma(P) = \mathrm{Im}\int_{P_o}^{P} dk$$

but now we have to specify the path, otherwise σ is not well defined. It follows that $d\sigma=\mathrm{Im}(dk)$ is a well defined 1-form.

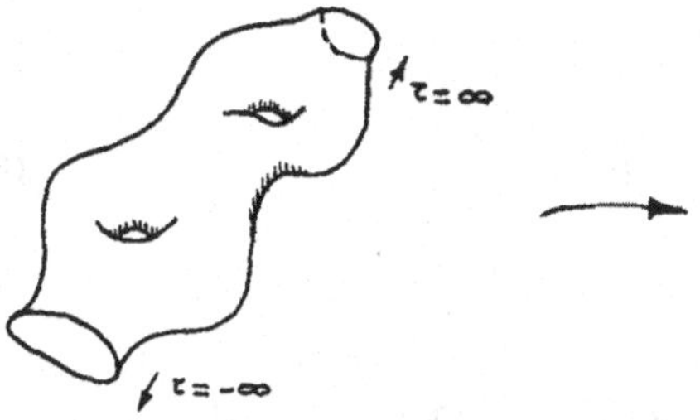

Fig.2a.

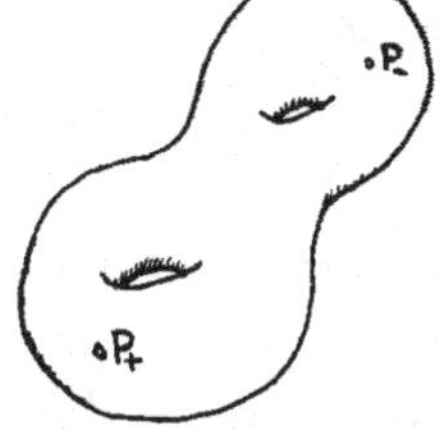

Fig.2b.

The string propagating along Σ will be represented by a one parameter family of contours C_τ defined as follows

$$C_\tau = \{ P \in \Sigma : \tau(P) = \tau \}$$

For $\tau \to \mp\infty$ the C_τ are small circles around $P_\pm$. As τ grows up, the string evolves with splitting and joinings until it reaches the point P_- (Fig.3).

Figure 3:

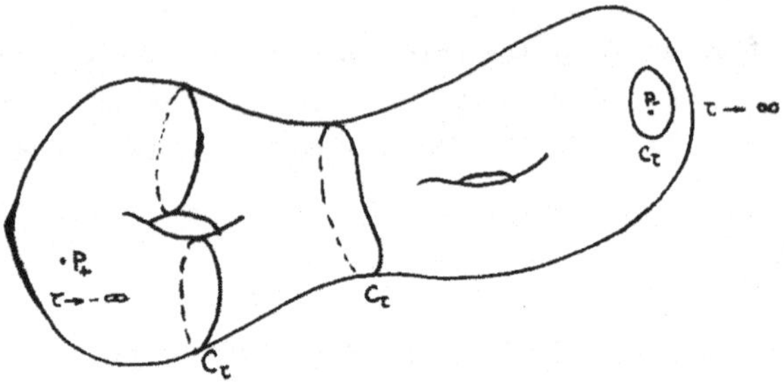

As we will see in the next section, the equation of motion of the matter field X^μ can be derived from a Hamiltonian and is written at any genus as

$$\partial\bar{\partial}X^\mu(Q) = 0 \quad , Q \neq P_+, P_- \tag{2.4}$$

(Note that if we had required that eq.(2.4) hold for any Q, then the only solution would have been the constant).

Eq.(2.4) implies that ∂X^μ ($\bar{\partial}X^\mu$) is holomorphic (antiholomorphic) everywhere except P_+ and P_-. At genus zero they can be therefore written as

$$\partial X^\mu(Q) = i/\sqrt{2} \sum_n \alpha_n^\mu z^{-n-1} dz \tag{2.5a}$$

$$\bar{\partial}X^\mu(Q) = i/\sqrt{2} \sum_n \bar{\alpha}_n^\mu \bar{z}^{-n-1} d\bar{z} \tag{2.5b}$$

Because we ask X^μ to be single-valued, one must require

$$\oint_{C_\tau} dX^\mu = 0 \tag{2.6}$$

This leads to

$$\alpha_0^\mu = \bar{\alpha}_0^\mu = -p^\mu/\sqrt{2} \tag{2.7}$$

X^μ is obtained by integration

$$X^\mu(z,\bar{z}) = \int^z dX^\mu = x^\mu - ip^\mu\tau - i/\sqrt{2} \sum_{n\neq 0} (\alpha_n^\mu z^{-n}/n + \bar{\alpha}_n^\mu \bar{z}^{-n}/n)$$

$$\text{where} \qquad \tau = 1/2 \log z\bar{z} = \int^z \mathrm{Re}(dz/z) \tag{2.8}$$

What about higher genus? Let us introduce

(ω_n) - basis for the space of meromorphic 1-forms which are holomorphic outside P_+ and P_-;

$(\bar{\omega}_n)$ - its complex conjugate.

Then ∂X^μ and $\bar{\partial} X^\mu$ can be expanded as follows

$$\partial X^\mu(Q) = i/\sqrt{2} \sum_n \alpha^\mu_n \, \omega_n(Q) \tag{2.9a}$$

$$\bar{\partial} X^\mu(Q) = i/\sqrt{2} \sum_n \bar{\alpha}^\mu_n \, \bar{\omega}_n(Q) \tag{2.9b}$$

In order to explicitely find the basis (ω_n) one makes use of the Riemann-Roch theorerm. This guarantees the existence and uniqueness of meromorphic 1-forms ω_n, holomorphic outside $P_\pm$, where they have the form

$$\begin{aligned} \omega_n(z_\pm) &= a^{(n)}_\pm \, z_\pm^{\mp n+g/2-1} \, (1 + o(z_\pm)) \, dz_\pm \quad , \quad |n|>g/2 \\ &= a^{(n)}_\pm \, z_\pm^{\mp n+g/2\mp{}^{1}_{0}} \, (1 + o(z_\pm)) \, dz_\pm \quad , \quad n \in I \\ \omega_{g/2} &= dk \qquad , \quad a^{(n)}_+ = 1 \quad \forall \, n \end{aligned} \tag{2.10}$$

where $I=[-g/2, g/2)$. Now the requirement of single-valuedness of X^μ implies the following relations

$$\oint_{C_\tau} dX^\mu = 0 \tag{2.11}$$

$$\oint_{\alpha_i} dX^\mu = 0 = \oint_{\beta_i} dX^\mu \tag{2.12}$$

By inserting the expansions (2.9), one finds that eq.(2.11) implies

$$\alpha^\mu_0 = \bar{\alpha}^\mu_0 = -p^\mu/\sqrt{2}$$

just as in the genus zero case, whereas eqs.(2.12) imply

$$\sum_n (\alpha^\mu_n a^i_n + \bar{\alpha}^\mu_n \bar{a}^i_n) = 0 \tag{2.13a}$$

$$\sum_n (\alpha^\mu_n b^i_n + \bar{\alpha}^\mu_n \bar{b}^i_n) = 0 \tag{2.13b}$$

where

$$a^i_n = \oint_{\alpha_i} \omega_n \qquad\qquad b^i_n = \oint_{\beta_i} \omega_n \tag{2.13c}$$

Eqs.(2.13a,b) can be written as

$$\sum_{n\in I} (\alpha^\mu_n a^i_n + \bar{\alpha}^\mu_n \bar{a}^i_n) = A^\mu_i \tag{2.14a}$$

$$\sum_{n\in I} (\alpha^\mu_n b^i_n + \bar{\alpha}^\mu_n \bar{b}^i_n) = B^\mu_i \tag{2.14b}$$

where

$$A^\mu_i = - \sum_{n\notin I} (\alpha^\mu_n a^i_n + \bar{\alpha}^\mu_n \bar{a}^i_n)$$

$$B^\mu_i = - \sum_{n\notin I} (\alpha^\mu_n b^i_n + \bar{\alpha}^\mu_n \bar{b}^i_n)$$

Eqs.(2.14) can be seen as a system of 2g equations with 2g unknowns. Note that for $n \in I$, (ω_n) is a basis of abelian differentials. Writing $\omega_n = \sum_{i=1}^{g} a^i_n \eta_i$, then eqs.(2.14a,b) become

$$\epsilon^\mu_i + \bar{\epsilon}^\mu_i = A^\mu_i \tag{2.15a}$$

$$\sum_{j=1}^{g} (\epsilon^\mu_j \Omega_{ji} + \bar{\epsilon}^\mu_j \bar{\Omega}_{ji}) = B^\mu_i \tag{2.15b}$$

where

$$\epsilon^\mu_i = \sum_{n\in I} \alpha^\mu_n a^i_n \tag{2.16}$$

In matrix notation

$$\begin{pmatrix} 1 & 1 \\ \Omega & \bar{\Omega} \end{pmatrix} \begin{pmatrix} \epsilon \\ \bar{\epsilon} \end{pmatrix} = \begin{pmatrix} A \\ B \end{pmatrix} \tag{2.17}$$

It follows $(\Omega=\Omega_1+i\Omega_2)$

$$\begin{pmatrix} \epsilon \\ \bar{\epsilon} \end{pmatrix} = \begin{pmatrix} \frac{i}{2}\Omega_2^{-1}\bar{\Omega} & -\frac{i}{2}\Omega_2^{-1} \\ -\frac{i}{2}\Omega_2^{-1}\Omega & \frac{i}{2}\Omega_2^{-1} \end{pmatrix} \begin{pmatrix} A \\ B \end{pmatrix} \tag{2.18}$$

As previously, X^μ is obtained by integration

$$X^\mu(Q) = \int_{Q_o}^{Q} dX^\mu = x^\mu - ip^\mu \tau(Q) + i/\sqrt{2} \sum_{n\neq g/2} (\alpha^\mu_n B_n(Q) + \bar{\alpha}^\mu_n \bar{B}_n(Q)) \tag{2.19}$$

where

$$B_n(Q) = \int_{Q_o}^{Q} \omega_n \quad ; \quad \tau(Q) = \mathrm{Re}\int_{Q_o}^{Q} dk$$

By using eqs.(2.16) and (2.18) the $\epsilon^\mu_i, \bar{\epsilon}^\mu_i$, $i=1,\dots,g$, are expressed in terms of

α^μ_n , $\bar\alpha^\mu_n$ with $|n|>g/2$. One obtains

$$\epsilon^\mu_i = -\sum_{|n|>g/2} (F_{in}\alpha^\mu_n + G_{in}\bar\alpha^\mu_n)$$

$$\bar\epsilon^\mu_i = -\sum_{|n|>g/2} (G_{in}\alpha^\mu_n + F_{in}\bar\alpha^\mu_n) \tag{2.20a}$$

with

$$F_{nj} = i/2 \sum_{i=1}^{g} ((\Omega_2^{-1}\bar\Omega)_{ji} a^i_n - (\Omega_2^{-1})_{ji} b^i_n) \tag{2.20b}$$

$$G_{nj} = i/2 \sum_{i=1}^{g} ((\Omega_2^{-1}\Omega)_{ji} \bar a^i_n - (\Omega_2^{-1})_{ji} \bar b^i_n) \tag{2.20c}$$

Inserting this into eq.(2.19), it follows

$$X^\mu(Q) = x^\mu - ip^\mu\tau(Q) + i/\sqrt{2} \sum_{|n|>g/2} (\alpha^\mu_n \phi_n(Q) + \bar\alpha^\mu_n \phi_n(Q)) \tag{2.21}$$

where $\phi_n(Q)$ are harmonic (single-valued) functions given by

$$\phi_n(Q) = \int_{Q_0}^{Q} (\omega_n - \sum_{j=1}^{g} (F_{nj}\eta_j + \bar G_{nj}\bar\eta_j)) \tag{2.22}$$

3.Hamiltonian formulation and equations of motion.

Let $t = t_{ab}d\sigma^a \otimes d\sigma^b$ be the energy-momentum tensor corresponding to some conformal field theory. We define the Hamiltonian in the standard way by ($\alpha'=1$)

$$H(\tau) = 1/2\pi \oint_{C_\tau} d\sigma\ t_{11} \tag{3.1}$$

(Recall $d\sigma=(dk-d\bar k)/2i$). The momentum is defined by

$$P(\tau) = 1/2\pi \oint_{C_\tau} d\sigma\ t_{12} \tag{3.2}$$

Taking into account the tracelessness and symmetry of t, the above equations become

$$H(\tau) = -1/2\pi \oint_{C_\tau} (t|e_\sigma) \tag{3.3a}$$

$$P(\tau) = 1/2\pi \oint_{C_\tau} (t|e_\tau) \tag{3.3b}$$

where e_τ, e_σ are defined through the following relations

$$e_\tau = e_k + e_{\bar{k}} \qquad , \; e_\sigma = i(e_k - e_{\bar{k}})$$

$$dk(e_k) = (e_k|dk) = 1 = (e_{\bar{k}}|d\bar{k}) \tag{3.4}$$

$$d\bar{k}(e_k) = (e_k|d\bar{k}) = 0 = (e_{\bar{k}}|dk)$$

Since $e_k \in K^{-1}$ it follows that it has degree (2-2g). dk has simple poles at P_+ and P_-, therefore e_k must have simple zeroes at these points. Thus the vector field e_k has 2g poles outside $P_\pm$, just on the zeroes of dk. These points correspond to those where the string splits or joins.

Let us now focus our attention on string theory. The phase space for the classical bosonic string theory at fixed τ is defined to be the space of functions X^μ and differentials P^μ, $\mu=0,\dots,D-1$, with the Poisson bracket

$$\{P^\mu(Q);X^\nu(Q')\} = -\eta^{\mu\nu}\,\Delta_\tau(Q,Q') \qquad ;, \; Q,Q' \in C_\tau \tag{3.5}$$

where $\eta^{\mu\nu}$ is the Minkowski metric with signature $(-1,1,\dots,1)$ and $\Delta_\tau(Q,Q')$ is the delta function over the contour C_τ (For any continuously differentiable function f on C_τ one has $f(Q)=\oint_{C_\tau} f(Q')\Delta_\tau(Q,Q')$).

In complex notation the energy-momentum tensor can be written as

$$t = T + \bar{T} \tag{3.6}$$

where

$$T = -1/4(dX + 2\pi P)^2 \tag{3.7a}$$

$$T = -1/4(dX - 2\pi P)^2 \tag{3.7b}$$

Inserting this expression in (3.3a) one obtains

$$H(\tau) = 1/2\pi \oint_{C_\tau} 1/2(dX^2 + 4\pi^2P^2|e_\sigma) \tag{3.8}$$

The equations of motion are easily obtained by imposing the canonical Poisson brackets

$$L_{e_\tau} X^\mu = i\{H;X^\mu\} = -2\pi i(P^\mu|e_\sigma) \tag{3.9a}$$

$$L_{e_\tau} P^\mu = i\{H;P^\mu\} = 1/2\pi i d(dX^\mu|e_\sigma) \tag{3.9b}$$

The first one tells us that

$$P^\mu = 1/2\pi(\partial - \bar{\partial})X^\mu \tag{3.10a}$$

Inserting this in eq.(3.9b) one obtains

$$L^2_{e_\tau} X^\mu = -L^2_{e_\sigma} X^\mu \tag{3.10b}$$

from where it follows $\bar{\partial}\partial X^\mu = 0$, as anticipated in section 2. In particular the components T and $\bar{T}$ of t can be written as

$$T = -\partial X\, \partial X \quad , \quad \bar{T} = -\bar{\partial}X\, \bar{\partial}X \tag{3.11}$$

Note that $\oint_{C_\tau} P^\mu = p^\mu$ (cf.eq.(2.21)), so p^μ is in fact the center of mass momentum of the string.

The next step is the quantization of the theory: the coefficients of the expansion (2.21) become second-quantized operators acting on a Fock space, whose commutation rules are to be derived from the canonical commutation relation

$$(P^\mu(Q);X^\nu(Q')) = -i\eta^{\mu\nu}\, \Delta_\tau(Q,Q') \quad ;, \; Q,Q' \in C_\tau \tag{3.12}$$

This leads to the following commutation rules for the α^μ_n, $\bar{\alpha}^\mu_n$ operators

$$[\alpha^\mu_n,\alpha^\nu_m] = \eta^{\mu\nu}\, \gamma_{nm} \quad , \quad [\bar{\alpha}^\mu_n,\bar{\alpha}^\nu_m] = \eta^{\mu\nu}\, \bar{\gamma}_{nm}$$

$$[\alpha^\mu_n,\bar{\alpha}^\nu_m] = 0 \quad , \quad [x^\mu,p^\nu] = i\eta^{\mu\nu} \tag{3.13}$$

where $\gamma_{nm} = 1/2\pi i \oint_{C_\tau} dA_n\, A_m$ ($\{A_n\}$ is a basis for the space of meromorphic functions, holomorphic outside $P_\pm$; for explicit expressions see [4,5,6]). At g=0 one has $A_n = z^n$, so $\gamma_{nm} = n\delta_{n+m}$.

The generalized Virasoro operators L_n, $\bar{L}_n$ are defined as the coefficients in the expansions of T and $\bar{T}$ respectively, that is

$$T(Q) = \sum_n L_n\, \Omega_n(Q) \tag{3.14a}$$

$$\bar{T}(Q) = \sum_n \bar{L}_n\, \bar{\Omega}_n(Q) \tag{3.14b}$$

where $\{\Omega_n\}$ ($\{\bar{\Omega}_n\}$) is a basis for the space of meromorphic quadratic differentials which are holomorphic (antiholomorphic) outside $P_\pm$. By using eqs.(2.9) and (3.11) we get the KN algebra

$$[L_i,L_j] = \sum_{s=-g_0}^{g_0} C^s_{ij}\, L_{i+j-s} + D\,\kappa_{ij} \quad ; \quad g_0=3/2g \tag{3.15}$$

where C^{s}_{ij} are the KN structure constants and κ_{ij} is the cocycle, whose dependence on the normal ordering consists just in the addition of trivial cocycles [4]. This algebra has the property that at g=0 reduces to the usual Virasoro algebra.

The Fock space is defined as usual to be the space generated by states made from the successive applications of α^{μ}_{n}, $\bar{\alpha}^{\mu}_{n}$ on a vacuum state. This is defined by

$$\alpha^{\mu}_{n}|0\rangle = \bar{\alpha}^{\mu}_{n}|0\rangle = 0 \quad \text{if } n\geq g/2$$

$$\langle 0|\alpha^{\mu}_{n} = \langle 0|\bar{\alpha}^{\mu}_{n} = 0 \quad \text{if } n<-g/2 \text{ or } n=g/2 \tag{3.16}$$

The normal ordered product is defined as

$$:\alpha^{\mu}_{n}\,\alpha^{\nu}_{m}: = \begin{cases} \alpha^{\mu}_{n}\alpha^{\nu}_{m} & \text{if } n<-g/2 \text{ or } m>g/2 \\ \alpha^{\nu}_{m}\alpha^{\mu}_{n} & \text{if } n>g/2 \text{ or } m<-g/2 \end{cases} \tag{3.17}$$

By using the relations (2.20), this induces the following normal ordering for the α^{μ}_{n}; $n \in I$:

$$:\alpha^{\mu}_{n}\alpha^{\nu}_{m}: = \alpha^{\mu}_{n}\alpha^{\nu}_{m} + \sum_{\substack{l>g/2 \\ k<-g/2}} \sum_{i,j=1}^{g} (a^{-1})_{in}(a^{-1})_{jm}(\gamma_{kl}F_{il}F_{jk} + \bar{\gamma}_{kl}G_{il}G_{jk}) \tag{3.18}$$

In terms of the $\bar{\alpha}^{\mu}_{n},\alpha^{\mu}_{n}$ operators the Hamiltonian and momentum take the form

$$H(\tau) = 1/2 \sum_{n,m} (l_{nm}(\tau):\alpha_{n}.\alpha_{m}: + \bar{l}_{nm}(\tau):\bar{\alpha}_{n}.\bar{\alpha}_{m}:) \tag{3.19a}$$

$$P(\tau) = i/2 \sum_{n,m} (l_{nm}(\tau):\alpha_{n}.\alpha_{m}: - \bar{l}_{nm}(\tau):\bar{\alpha}_{n}.\bar{\alpha}_{m}:) \tag{3.19b}$$

where

$$l_{nm}(\tau) = 1/2\pi i\oint_{C_{\tau}} (e_{k}|\omega_{n})\omega_{m} = l_{mn}(\tau)$$

Note that at g=0 they reduce to the known expressions $H=L_{o}+\bar{L}_{o}$; $P=i(L_{o}-\bar{L}_{o})$.

As a final remark, let us stress that H and P depend on time. This is due to the 2g poles of the vector fields e_{k}, e^{-}_{k} . The variation of $l_{nm}(\tau)$ with τ is however very simple. It is like a step function in the sense that it remains constant until it reaches a splitting or joining of the C_{τ} (because the integrand picks a pole from e_{k}), where it changes value by a discrete quantity.

4. Vertex operators and scattering amplitudes.

Let $A^M[\Sigma]$ denote the amplitude of an M-particle scattering process performed on an equivalence class of Riemann surface $[\Sigma]$. The total scattering amplitude will be given by the sum of $A^M[\Sigma]$ over all equivalences classes and all topologies with a certain weight. Here we will be only concerned with the definition of $A^M[\Sigma]$.

In string theory, a scattering amplitude is given by the vacuum expectation value of the time-ordered product of suitable vertex operators V_i; $i=1,\ldots,M$

$$A^M[\Sigma] = \langle 0|T(V_1\ldots V_M)|0\rangle \tag{4.1}$$

The right definition of the vertex operators can be found by looking at the genus zero case. In general we expect something of the form

$$V = \int_\Sigma W(Q;p^\mu) \tag{4.2}$$

where $W(Q;p^\mu)$ is a (1,1) form.

At g=0 one has for the M tachyon scattering amplitude

$$A^M_{g=0} = \int_\Sigma d\mu_M(z) \langle 0|T(e^{ip_1\cdot X(z_1,\bar{z}_1)}\ldots e^{ip_M\cdot X(z_M,\bar{z}_M)})|0\rangle \tag{4.3}$$

where

$$d\mu_M(z) = |z_A-z_B|^2|z_A-z_C|^2|z_B-z_C|^2\delta^2(z_A-z_A^0)\delta^2(z_B-z_B^0)\delta^2(z_C-z_C^0)\prod_{i=1}^M \frac{(dz_i\wedge d\bar{z}_i)}{z_i\,\bar{z}_i}$$

At genus g>1 there is no residual symmetry, so we expect that near $P_\pm$ one has

$$d\mu_M(z) \propto \prod_{i=1}^M \frac{(dz_i\wedge d\bar{z}_i)}{z_i\,\bar{z}_i}$$

Riemann-Roch guarantees that up to the addition of holomorphic terms the only one differential with this behaviour near $P_\pm$ and which have no other poles anywhere on Σ is dk, and similarly for $d\bar{k}$. So we are led to the definition

$$W_{tachyon} = dk\wedge d\bar{k}\; e^{ip.X}$$

$$= (-2i)d\tau\wedge d\sigma\; e^{ip.X} \tag{4.4}$$

Similarly for the massless level, we define the graviton vertex as

$$V_{grav} = \xi_{\mu\nu}(p)\int_\Sigma \partial X^\mu\wedge\bar{\partial}X^\nu\; e^{ip.X} \tag{4.5}$$

where $\xi_{\mu\nu}(p)=\xi_{\nu\mu}(p)$ obeys $\xi_{\mu\nu}p^{\nu}=\xi^{\mu}{}_{\mu}=p^2=0$. If $\xi_{\mu\nu}(p)=-\xi_{\nu\mu}(p)$, this vertex represents the antisymmetric particle. The dilaton is obtained by the replacement $\xi_{\mu\nu} \rightarrow \eta_{\mu\nu}$. As explained in ref.[7], for successive levels one should expect vertices of the form

$$V_{Y_n} = \xi_{\mu_1 \ldots \mu_{2n}}(p) \int_{\Sigma} \partial X^{\mu_1} \wedge \bar{\partial} X^{\mu_2} \, (e_k | \partial X^{\mu_3})(e_{\bar{k}} | \bar{\partial} X^{\mu_4}) \ldots (e_{\bar{k}} | \bar{\partial} X^{\mu_{2n}}) \; e^{ip.X} \qquad (4.6)$$

and other vertices with higher derivatives as well (for a study of vertex operators in the path integral approach and in conformal theories see respectively [8,9]).

In general, the polarization tensors as well as the momentum p^{μ} must satisfy certain constraints (which lead to the mass and characterization of the type of particle represented by the operator vertex) coming from the requirement for W to have dimension (1,1), in the sense that

$$[L_n, W] = L_{e_n} W \qquad (4.7a)$$

$$[\bar{L}_n, W] = L_{\bar{e}_n} W \qquad (4.7b)$$

These relations are easily verified for the massless level. For other vertices the normal ordering (eqs.(3.17,18)) plays a crucial role and thus the computation of these commutators is more involved. For instance, in the case of the tachyon one obtains

$$[L_r, :e^{ip.X}: dk \wedge d\bar{k}] = (e_r | :\partial \, e^{ip.X}:) dk \wedge d\bar{k} + p^2/4 :e^{ip.X}: \kappa_r \, dk \wedge d\bar{k} \qquad (4.8)$$

where

$$\kappa_r = \Big(\sum_{\substack{m \\ n>g>2}} - \sum_{\substack{n \\ m \le -g/2}} \Big) l^r_{nm} A_n A_m + \sum_{\substack{m \\ n \in I}} l^r_{nm} A_m \sum_{l>g/2} \sum_{i=1}^{g} (a^{-1})_{in} (F_{il} A_l + G_{il} \bar{A}_l)$$

$$- \sum_{\substack{n \\ m \in I}} l^r_{nm} A_n \sum_{l \le -g/2} \sum_{i=1}^{g} (a^{-1})_{in} (F_{il} A_l + G_{il} \bar{A}_l) + (e_r | dk) \qquad (4.9)$$

A further study of the sum in (4.9) is needed to verify whether (4.8) reduces to (4.7a).

The commutators (4.7) are important because they play a crucial role in proving the decoupling of ghosts (negative norm states) of the amplitude (4.1) [7].

5. Correlation functions for the matter field.

It is remarkable that computations at arbitrary genus can be explicitely performed and well-known results naturally arise. Here we will calculate $<\partial X^{\mu}\bar{\partial}X^{\nu}>$, leaving other correlation functions as well as scattering amplitudes to be reported elsewhere. This correlation function, in this context, is given by

$$<0|T(\partial X^{\mu}(Q)\bar{\partial}X^{\nu}(Q'))|0> = f^{\mu\nu}(Q,Q')\theta(\tau_{Q}-\tau_{Q'}) + g^{\mu\nu}(Q,Q')\theta(\tau_{Q'}-\tau_{Q}) \tag{5.1}$$

where $\quad f^{\mu\nu}(Q,Q') = <0|\partial X^{\mu}(Q)\bar{\partial}X^{\nu}(Q')|0> \quad , \quad g^{\mu\nu}(Q,Q') = <0|\bar{\partial}X^{\nu}(Q')\partial X^{\mu}(Q)|0>$

By inserting the expansions (2.9a,b) and eq.(2.20) one obtains

$$f^{\mu\nu}(Q,Q') = \eta^{\mu\nu} \sum_{i,j=1}^{g} C_{ij}\ \eta_i(Q)\ \bar{\eta}_j(Q') = g^{\mu\nu}(Q,Q') \tag{5.2}$$

where

$$\eta^{\mu\nu}C_{ij} = -1/2 <0|\epsilon^{\mu}_{i}\ \epsilon^{\nu}_{j}\ |0>$$

$$= -1/2\ \eta^{\mu\nu} \sum_{n>g/2}\ \sum_{m<-g/2} (\ \gamma_{nm}F_{in}\bar{G}_{jm} + \bar{\gamma}_{nm}\bar{F}_{jm}\ G_{in}) \tag{5.3}$$

In order to calculate C_{ij} we use the relations

$$\sum_{n} \gamma_{nm}\ F_{im} = \sum_{n} \gamma_{nm}\ \bar{G}_{im} = 0 \tag{5.4}$$

where one has

$$F_{in} = a^{i}_{n}\ ,\quad G_{in} = 0 \quad ;\qquad n \in I \tag{5.5}$$

Therefore C_{ij} can be rewritten as

$$C_{ij} = -\sum_{n\in I} a^{i}_{n}\ D_{nj} \tag{5.6}$$

with

$$D_{nj} = 1/2 \sum_{m>g/2} \gamma_{nm}\ \bar{G}_{jm} \tag{5.7}$$

By using eq.(2.20c) it follows

$$D_{nj} = i/4 \sum_{i=1}^{g} (\Omega_2^{-1})_{ji}\ K_{ni} \tag{5.8}$$

where

$$K_{ni} = \sum_{m>g/2} \gamma_{nm}\ (\ b^{i}_{m} - \sum_{j=1}^{g} \Omega_{ij}\ a^{j}_{m}\ ) \tag{5.9}$$

Now we use the following identity (for reasons of space, the details of the proof will be given elsewhere)

$$b^i_m - \sum_{j=1}^{g} \Omega_{ij}\, a^j_m = \oint_{C_+} I_i\, \omega_m \quad , \quad m>g/2 \tag{5.10}$$

where $I_i(Q)=\int^Q \eta_i$, and C_+ is a small contour around P_+. It follows

$$K_{ni} = \oint_{C_+} I_i \sum_{m>g/2} \gamma_{nm}\, \omega_m = \oint_{C_+} I_i\, dA_m = -2\pi i\, (a^{-1})^i_n \tag{5.11}$$

where we have used the fact that ω_n, $n<g/2$ are holomorphic in P_+ . Now from (5.8) we have

$$D_{nj} = \pi/2 \sum_{l=1}^{g} (\Omega_2^{-1})_{jl}\, (a^{-1})^l_n \tag{5.12}$$

Inserting this result into (5.6) we get

$$C_{ij} = -\pi/2\, (\Omega_2^{-1})_{ij} \tag{5.13}$$

Thus we finally obtain

$$<0|T(\partial X^\mu(Q)\bar\partial X^\nu(Q'))|0> = \eta^{\mu\nu}\, \pi \sum_{i,j=1}^{g} (\Omega_2^{-1})_{ij}\, \eta_i(Q)\, \bar\eta_j(Q') \tag{5.14}$$

which coincides with the well-known result quoted in the literature, computed by other methods [10].

References

[1] M.Green, J. Schwarz and E. Witten: "Superstring Theory",vol.I and II, Cambridge University Press, 1987; and references therein.

[2] A.M.Polyakov, Phys. Lett.103B (1981),207 and 211;
D.Friedan, in Les Houches 1982, Recent advances in Field Theory and Statistical Physics, eds. J.B.Zuber and R.Stora (North Holland,1984);
O.Alvarez,Nucl.Phys.B216 (1983)125.

[3] L.Alvarez-Gaume, C.Gomez, G.Moore and C.Vafa: "Strings in the operator formalism", BUHEP-87/51;
L.Alvarez-gaume, C.Gomez, P.Nelson, G.Sierra and C.Vafa, preprint BUHEP-88/11.

[4] I.M.Krichever and S.P.Novikov, Funk. Anal.i.Pril.21 No.4 (1987),47.

[5] I.M.Krichever and S.P.Novikov, Funk. Anal.i.Pril.21 No.2 (1987),46.

[6] L.Bonora, A.Lugo, M.Matone and J.Russo: "A global operator formalism on

higher genus Riemann surfaces:b-c systems", preprint SISSA 67/88/EP.
[7] A.Lugo and J.Russo: "Hamiltonian formulation and scattering amplitudes in string theory at genus g", preprint SISSA 83/88/EP.
[8] S.Weinberg,Phys.Lett.156B (1985) 309.
[9] R.Sasaki and I.Yamanaka,Phys.Lett.165B (1985) 283.
[10] E.Verlinde and H.Verlinde, Nucl.Phys.B288 (1987), 357;
H.Sonoda, Phys.Lett.178B (1986),390;
M.Bonini and R.Iengo, Int.J.Mod.Phys.A3 (1988), 841.

CONFORMAL FIELD THEORY, REAL WEIGHT DIFFERENTIALS AND KDV EQUATION IN HIGHER GENUS

MARCO MATONE

International School for Advanced Studies
Strada Costiera, 11 - 341014 Trieste, Italy
and
INFN, Sezione di Trieste

1. Introduction

In the last few years it has been recognized that two - dimensional conformal field theori on higher genus Riemann surfaces play a fundamental role in understanding the structure string theory. Recently Krichever and Novikov proposed a formalism for studying conform field theories in arbitrary genus [1]. One of the essential aspects of their approach is t introduction of bases for meromorphic differentials that can be seen as the generalization higher genus of the monomials z^n on the sphere. This fact allows us to define an operat formalism which closely resembles the operator formalism on the sphere [2].

Another result of KN's approach is the generalization of the Virasoro algebra to Riemann surface Σ: the KN algebra. It turns out that the representations of KN a Virasoro algebras are inequivalent [3]. The KN algebra plays a key role in showing that t critical dimension for the bosonic and supersymmetric strings are equal to the genus ze case [4]. In [5] the generalization of the Sugawara construction to higher genus was shown

In this paper we discuss some aspects concerning the generalization of the KN's bas to real weight differentials, in particular in section 2 we write these bases in terms of the functions and in section 3 we use them to define an operator formalism for a real weight b system, which is important for the study of conformal field theories in arbitrary genus [6].

In section 4 we formulate the KdV equation in higher genus showing its connection wi the KN algebra [7], this is a generalization of the relation between the second hamiltoni structure of the KdV equation on the cylinder and the Virasoro algebra [8]. In particular way to covariantize the expressions of the KdV formalism in higher genus is provided. Amo other things, we get an explicit form for the projective connection entering in the definiti of the cocycle of the central extension of KN algebra. Finally in the appendix we fix t notations about theta functions.

2. Real weight differentials

The real weight differentials on a Riemann surface are in general well defined only o particular covering, however for brevity we will call them λ - differentials. Let Σ be a comp genus g Riemann surface and $P_+, P_- \in \Sigma$ two distinguished points in general position t for $g = 0$ can be identified with $0, \infty$. In the following, $z_\pm$ will denote local coordina vanishing at $P_\pm$. For $\lambda \in \mathbf{R}$ we define a λ - differential holomorphic outside $P_\pm$ with following behaviour in a neighborhood of these points:

$$f_j^{(\lambda)}(z_\pm|l) = a_j^{(\lambda)\pm}(l) z_\pm^{\pm j - s(\lambda)}(1 + O(z_\pm))(dz_\pm)^\lambda, \qquad s(\lambda) = \frac{g}{2} - \lambda(g-1), \qquad (2$$

where $j \in \mathbf{Z} + P(l,\lambda)$, $P(l,\lambda) = s(\lambda) + \lambda(l-1)$, $l \in \mathbf{Z}$ and $a_j^{(\lambda)\pm}(l)$ are (not independe constants. For $|j| \le \frac{g}{2}$ there are some modifications discussed below. When $\lambda = \frac{m}{n}$, with ad n relatively prime numbers, there are only n distinct sectors, i.e. l =1,...,n. When z_+ ans a circle around P_+ not including P_-, $f_j^{(\lambda)}(z|l)$ is multivalued:

$$f_j^{(\lambda)}(z_\pm|l) \to e^{2\pi i\lambda(l-g\pm g)} f_j^{(\lambda)}(z_\pm|l). \tag{2.2}$$

otice that the phase factor contribution coming from $(dz_-)^\lambda$ is the inverse of the phase ue to the $(dz_+)^\lambda$ term. When $\lambda \in \mathbb{Z} + \frac{1}{2}$ the Neveu - Schwarz (l =1) and the Ramond =2) sectors are recovered. To understand this point we recall that the coordinate z_+ on e (punctured) sphere is related to the cylinder coordinate $w = \tau + i\sigma$ by the conformal map $= lnz_+$, so that for $\tau = const$ we have that $(dz_+)^\lambda = const\, e^{i\lambda\sigma}(d\sigma)^\lambda$.

Actually $f_j^{(\lambda)}(z|l)$ is a well defined λ - differential only if $2\lambda(g-1)$ is an integer number d $l = 1$, its multivaluedness being due only at the $(dz_\pm)^\lambda$ term in (2.1):

$$f_j^{(\lambda)}(z_\pm|1) \to e^{\pm 2\pi i\lambda} f_j^{(\lambda)}(z_\pm|1). \tag{2.3}$$

this case and when $\lambda > 1$, $g > 1$, the λ - form $f_j^{(\lambda)}(z|1)$ is holomorphic for $s(\lambda) \le j \le$ $s(\lambda)$, therefore, being $-2s(\lambda) + 1 = (2\lambda - 1)(g-1)$, eq.(2.1) gives all the zero modes of the auchy - Riemann operator $\bar\partial$ coupled to λ - differentials.

Due to the Riemann - Roch theorem the eq.(2.1) does not work in the following cases his follows also from the application of the Riemann vanishing theorem to the explicit pression of the eq.(2.1) given in eq.(2.15))

$\forall g$ *for* $|j| \le \frac{g}{2}$ *and* $\lambda = 0, 1$;
$\forall g$ *for* $|j| = \frac{1}{2}$ *and* $\lambda = \frac{1}{2}$, $l = 1$ *with odd spin structure*;
$g = 1$ *for* $|j| = \frac{1}{2}$ *and* $\lambda \in \mathbb{Z}$;
$g = 1$ *for* $|j| = \frac{1}{2}$ *and* $\lambda \in \mathbb{Z} + \frac{1}{2}$, $l = 1$.

en we define

$$f_j^{(0)}(z_\pm) = a_j^{(0)\pm} z_\pm^{\pm j - \frac{g}{2} - \frac{(1\mp 1)}{2}}(1 + O(z_\pm)), \qquad -\frac{g}{2} \le j \le \frac{g}{2} - 1, \tag{2.4}$$

d $f_{\frac{g}{2}}^{(0)}(z) = 1$. For $\lambda = 1$ we put

$$f_j^{(1)}(z_\pm) = a_j^{(1)\pm} z_\pm^{\pm j + \frac{g}{2} - \frac{(1\pm 1)}{2}}(1 + O(z_\pm))dz_\pm, \qquad -\frac{g}{2} + 1 \le j \le \frac{g}{2}, \tag{2.5}$$

d choose $f_{-\frac{g}{2}}^{(1)}(z)$ to be the third kind abelian differential with simple poles at $P_\pm$ with idue ± 1, normalized in such a way that its periods are purely imaginary, i.e. $Re \oint_{a_i} f_{-\frac{g}{2}}^{(1)}(z)$ $Re \oint_{b_i} f_{-\frac{g}{2}}^{(1)}(z) = 0$. For $g = 1$ we define

$$f_j^{(\lambda)}(z) = f_j^{(0)}(z)(f_{\frac{1}{2}}^{(1)}(z))^\lambda, \quad \lambda \in Z, \tag{2.6}$$

$$f_j^{(\lambda)}(z|1) = f_j^{(0)}(z)(f_{\frac{1}{2}}^{(1)}(z))^\lambda, \quad \lambda \in Z + \frac{1}{2}, \tag{2.7}$$

ere the spin structure of $(f_{\frac{1}{2}}^{(1)}(z))^{\frac{1}{2}}$ is taken odd. Finally, for $\lambda = \frac{1}{2}$, l =1 and odd spin ucture:

$$f_{-\frac{1}{2}}^{(\frac{1}{2})}(z_\pm|1) = a_{-\frac{1}{2}}^{(\frac{1}{2})\pm}(1) z_\pm^{-1}(1 + O(z_\pm))(dz_\pm)^{\frac{1}{2}}, \tag{2.8}$$

$$f^{(\frac{1}{2})}_{\frac{1}{2}}(z_\pm|1) = a^{(\frac{1}{2})\pm}_{\frac{1}{2}}(1)(1+O(z_\pm))(dz_\pm)^{\frac{1}{2}}. \quad (2.9$$

The Riemann - Roch theorem guarantees the existence and uniqueness of $f_j^{(\lambda)}(z|l)$ u to the multiplicative constant $a_j^{(\lambda)+}(l)$ ($a_j^{(\lambda)-}(l)$ is completely determined from the $a_j^{(\lambda)+}($ choice); we will show its existence by explicit construction. The uniqueness follows fro the fact that given two λ - differentials satisfying eq.(2.1), their quotient is a meromorph function with g poles in general position, however, by the Noether gap theorem [9], th function is a constant.

On Σ the euclidean time can be defined to be the harmonic function

$$\tau = Re\int_{Q_0}^{Q} f^{(1)}_{-\frac{g}{2}}(z). \quad (2.1$$

In this way the level lines

$$C_\tau = \{Q\in\ \Sigma|Re\int_{Q_0}^{Q} f^{(1)}_{-\frac{g}{2}}(z) = \tau\}, \quad (2.1$$

are identified as the configurations of the string at the time τ. For $\tau\to\pm\infty$,C_τ becon small circles around $P_\mp$. The definition of the contours C_τ allows us to define the dual $f_j^{(\lambda)}(z|l)$ by

$$\frac{1}{2\pi i}\oint_{C_\tau} f_i^{(\lambda)}(z|l) f^j_{(1-\lambda)}(z|l) = \delta^j_i, \quad (2.1$$

where the explicit form of the dual $f^j_{(1-\lambda)}(z|l)$ will be shown later.

Let us start now with the construction of the λ - differential $f_j^{(\lambda)}(z|l)$ in terms of the functions and prime forms. The expression

$$\frac{E(z,P_+)^{j+s(\lambda)-2\lambda}}{E(z,P_-)^{j+s(\lambda)}},$$

is a (multivalued) λ - differential with $2s(\lambda)-2\lambda = g(1-2\lambda)$ zeroes in P_+ more than $f_j^{(\lambda)}(z$ Since by the Riemann vanishing theorem [10] $div\theta(P-gP_++\Delta) = gP_+$, we put

$$f_j^{(\lambda)}(z|l) = \frac{E(z,P_+)^{j+s(\lambda)-2\lambda}}{E(z,P_-)^{j+s(\lambda)}}\frac{\theta(z+(j-s(\lambda))P_+-(j+s(\lambda))P_-+(1-2\lambda)\Delta)}{\theta^{1-2\lambda}(z-gP_++\Delta)}, \quad (2.1$$

where the Jacobi map of the points in the theta functions argument is understood. T θ - function at the numerator guarantees the singlevaluedness of $f_j^{(\lambda)}(z|l)$, on the other ha since it has g - zeroes, it follows that the degree of $f_j^{(\lambda)}(z|l)$ is precisely $2\lambda(g-1)$. To ma manifest the divisor of the right hand side of eq.(2.13) note that (for semplicity we consi the case $\lambda\in\mathbf{Z}$)

$$2\lambda\Delta = [K^\lambda] = [(j+s(\lambda))P_+-(j+s(\lambda))P_-+\Sigma_{i=1}^g P_i],$$

so that by Abel theorem it follows that $\theta(z)$ has the same zeroes of $\theta(z+2\lambda\Delta-[K^\lambda$ therefore for the θ - function in (2.13) we have

$$div\theta(z+(j-s(\lambda))P_+-(j+s(\lambda))P_-+(1-2\lambda)\Delta) = div\theta(z-\Sigma_{i=1}^g P_i+\Delta), \quad (2.$$

here the zeroes $P_1, ..., P_g$ are precisely those fixed by the conditions (2.1).

Using the definition of the σ - differential (see eq.(A.11)) and inserting the theta characteristics, eq.(2.13) is generalized to the λ - differential with λ - *structure* $[^{\alpha}_{\beta}]$, i.e.

$$f_j^{(\lambda)}(z|l) = \frac{\theta[^{\alpha}_{\beta}](z + (j - s(\lambda))P_+ - (j + s(\lambda))P_- + (1 - 2\lambda)\Delta)\sigma^{2\lambda-1}(z)}{E(z, P_+)^{-j+s(\lambda)}E(z, P_-)^{j+s(\lambda)}}, \tag{2.15}$$

here $j \in \mathbf{Z} + P(l, \lambda)$, $P(l, \lambda) = s(\lambda) + \lambda(l - 1)$, $l \in \mathbf{Z}$. If $\lambda \notin \mathbf{Q}$, then $\alpha_i, \beta_i \in [0, 1]$ whereas $\lambda = \frac{m}{n}$ then $\alpha_i, \beta_i \in \{0, \frac{1}{n}, ..., \frac{n-1}{n}\}$, so that there are n^{2g} λ - *structures*, n for each one the a_i and b_i homology cycles. The dual of $f_j^{(\lambda)}(z|l)$ is (up to a multiplicative constant etermined from eq.(2.12))

$$f^j_{(1-\lambda)}(z|l) = \frac{\theta[^{-\alpha}_{-\beta}](z - (j - s(\lambda) + 1)P_+ + (j + s(\lambda) - 1)P_- + (2\lambda - 1)\Delta)}{E(z, P_+)^{j-s(\lambda)+1}E(z, P_-)^{-j-s(\lambda)+1}\sigma^{2\lambda-1}(z)}. \tag{2.16}$$

he multivaluedness of $f^j_{(1-\lambda)}(z|l)$ is given by (2.2) with λ replaced by $1 - \lambda$, so that the tegrand in eq.(2.12) is a well defined 1 - differential.

The explicit form of the differentials in eqs.(2.4 - 5) is

$$f_j^{(0)}(z) = \frac{E(z, R)\theta(z + (j - \frac{g}{2})P_+ + R - (j + \frac{g}{2} + 1)P_+ + \Delta)}{E(z, P_+)^{-j+\frac{g}{2}}E(z, P_-)^{j+\frac{g}{2}+1}\sigma(z)}, \qquad -\frac{g}{2} \le j \le \frac{g}{2} - 1, \tag{2.17}$$

$$f_j^{(1)}(z) = \frac{\sigma(z)\theta(z + (j + \frac{g}{2} - 1)P_+ - R - (j - \frac{g}{2})P_- - \Delta)}{E(z, P_+)^{-j-\frac{g}{2}+1}E(z, P_-)^{j-\frac{g}{2}}E(z, R)}, \qquad -\frac{g}{2} + 1 \le j \le \frac{g}{2}, \tag{2.18}$$

here the point $R \in \Sigma$ is arbitrary; its presence in (2.17) reflects the fact that the meromorhic function defined in (2.4) is unique up to the addition of a constant. In eq.(2.18) the le in R is cancelled by the zero of the θ - function; indeed, being $\theta(-z) = \theta(z)$, from the iemann vanishing theorem we have for $-\frac{g}{2} + 1 \le j \le \frac{g}{2}$

$$\theta((j + \frac{g}{2} - 1)P_+ - (j - \frac{g}{2})P_- - \Delta) = 0.$$

ote that this result allows to identify the differential $f_{\frac{1}{2}}^{(1)}(z)$ in eq.(2.6) with $\sigma^2(z)$.

For $j = -\frac{g}{2}$ the 1 - differential is

$$f_{-\frac{g}{2}}^{(1)}(z) = d\,ln\frac{E(z, P_+)}{E(z, P_-)} - 2\pi i \sum_{j,k=1}^{g} Im\left(\int_{P_-}^{P_+} \omega_k\right)(Im\Omega)^{-1}_{jk}\omega_k(z). \tag{2.19}$$

nally the conditions (2.8 - 9) give

$$f_{-\frac{1}{2}}^{(\frac{1}{2})}(z|1) = \frac{E(z, S)}{E(z, P_+)E(z, P_-)}\theta[^{\alpha}_{\beta}](z + S - P_+ - P_-), \tag{2.20}$$

$$f_{\frac{1}{2}}^{(\frac{1}{2})}(z|1) = \frac{\theta[^{\alpha}_{\beta}](z - P_-)}{E(z, P_-)}, \tag{2.21}$$

ere the θ - characteristics are odd. The presence in eq.(2.20) of the arbitrary point $S \in \Sigma$ due to the fact that eq.(2.8) does not fix uniquely $f_{-\frac{1}{2}}^{(\frac{1}{2})}(z|1)$. Note that $f_{\frac{1}{2}}^{(\frac{1}{2})}(z|1)$ as defined in (2.21) is equal to the expression (2.13) with $j=\frac{1}{2}$, $\lambda=\frac{1}{2}$, $l=1$. This means th in the framework of θ - functions theory the modification (2.9) to eq.(2.1) is automatical taken into account. Moreover since $P_-\in div\theta[^\alpha_\beta](z-P_-)$, $f^{(\frac{1}{2})}_{\frac{1}{2}}(z|1)$ has no poles, it is t zero - mode associated to the odd spin structure $[^\alpha_\beta]$, in particular using the definition of t prime form, we have

$$f^{(\frac{1}{2})}_{\frac{1}{2}}(z|1)=h(P_-)h(z), \qquad (2.2$$

so that we can identify $f^{(\frac{1}{2})}_{\frac{1}{2}}(z|1)$ with the $\frac{1}{2}$ - differential which appears in eq.(2.7).

We conclude this section giving the expression of the covariant delta function for λ differentials in the sector l:

$$\Delta^{(l)}(z,w)=\sum_j f^{(\lambda)}_j(z|l)f^j_{(1-\lambda)}(w|l), \qquad (2.2$$

that is, if $g(z)$ is a smooth λ - differential with the multivaluedness given in eq.(2.2), we ha

$$g(z)=\oint_{C_\tau}\Delta^{(l)}(z,w)g(w). \qquad (2.2$$

3. The b - c system in the operator formalism

The chiral b - c system is a first order system, the action of which is

$$S=\int_\Sigma b\bar{\partial}c, \qquad (3$$

the fields b and c being sections of K^λ and $K^{1-\lambda}$ respectively. We quantize this syst considering for simplicity the anticommuting case

$$\{b^i,c_j\}=\delta^i_j, \qquad \{b^i,b^j\}=0=\{c_i,c_j\}, \qquad (3$$

where $j\in\mathbf{Z}+P(l,\lambda)$, $P(l,\lambda)=s(\lambda)+\lambda(l-1)$, $l\in\mathbf{Z}$, $\lambda\in\mathbf{R}$. In the operator formali these fields are expanded in a basis of λ - differentials in such a way that the equations motion $\bar{\partial}b=0$ and $\bar{\partial}c=0$ are satisfied everywhere except at $P_\pm$. We use the different defined in section 2 as basis to expand the b and c fields

$$b(z|l)=\sum_j b^j f^{(\lambda)}_j(z|l), \qquad c(w|l)=\sum_j c_j f^j_{(1-\lambda)}(z|l). \qquad ($$

We define the vacuum $|0>$ associated to Σ generalizing the $\lambda\in\frac{\mathbf{Z}}{2}$ case discussed in ref.[2 the $\lambda\in\mathbf{R}$ case (for $\lambda=0,1$ *or* $g=1$ there are some modifications [2]):

$$\begin{aligned} &b^j|0>=<0|c_j=0, \qquad for\ j\le Q(l,\lambda)-1;\\ &c_j|0>=<0|b^j=0, \qquad for\ j\ge Q(l,\lambda), \end{aligned} \qquad ($$

where $Q(l,\lambda)=s(\lambda)+\overline{\lambda(l-1)}$ (the bar denotes the non integer part of $\lambda(l-1)$). propagator is

$$S^{(l)}(z,w)\equiv<0|T\{b(z|l)c(w|l)\}|0>=\begin{cases}<0|b(z|l)c(w|l)|0>, & \text{if } \tau_z>\tau_w;\\ -<0|c(w|l)b(z|l)|0>, & \text{if } \tau_w>\tau_z\,.\end{cases} \tag{3.5}$$

nserting the expansions (3.2) in eq.(3.5) and defining $<0|0>=1$ (as we will see this definition s consistent) we obtain

$$S^{(l)}(z,w)=\begin{cases}\sum_{j\leq Q(l,\lambda)-1} f_j^{(\lambda)}(z|l)f^j_{(1-\lambda)}(w|l), & \text{if } \tau_z>\tau_w;\\ -\sum_{j\geq Q(l,\lambda)} f_j^{(\lambda)}(z|l)f^j_{(1-\lambda)}(w|l), & \text{if } \tau_w>\tau_z.\end{cases} \tag{3.6}$$

'o evaluate $S^{(l)}(z,w)$ we look at the behaviour of the right hand side of eq.(3.6) in a neighorhood of $P_\pm$ and in the limit $z\to w$ (recall that $f_j^{(\lambda)}(z|l)$ and $f^j_{(1-\lambda)}(z|l)$ are defined up to multiplicative constant, determined from eq.(2.12), that we omit for brevity):

$$\begin{aligned} z\to P_+ &\Longrightarrow S^{(l)}(z,w)\sim(z-P_+)^{\overline{\lambda(l-1)}}\,;\\ z\to P_- &\Longrightarrow S^{(l)}(z,w)\sim(z-P_-)^{-2s(\lambda)+1-\overline{\lambda(l-1)}}\,;\\ w\to P_+ &\Longrightarrow S^{(l)}(z,w)\sim(w-P_+)^{-\overline{\lambda(l-1)}}\,;\\ w\to P_- &\Longrightarrow S^{(l)}(z,w)\sim(w-P_-)^{2s(\lambda)-1+\overline{\lambda(l-1)}}\,;\\ z\to w &\Longrightarrow S^{(l)}(z,w)\sim(z-w)^{-1}\,. \end{aligned} \tag{3.7}$$

ote that in the first and forth cases $\tau_w>\tau_z$, whereas in the second and third cases $\tau_z>\tau_w$. ince from (2.12) and (2.16) it follows that if $b(z|l)$ has λ - *structure* $[{\alpha \atop \beta}]$ then $c(z|l)$ has $.-\lambda)$ - *structure* $[{-\alpha \atop -\beta}]$, the conditions (3.7) give (for $\lambda\neq 0,1$ *and* $g\neq 1$)

$$S^{(l)}(z,w)=\frac{1}{E(z,w)}\left(\frac{E(z,P_-)}{E(w,P_-)}\right)^{-2s(\lambda)+1-\overline{\lambda(l-1)}}\left(\frac{E(z,P_+)}{E(w,P_+)}\right)^{\overline{\lambda(l-1)}}\cdot$$

$$\cdot\left(\frac{\sigma(z)}{\sigma(w)}\right)^{2\lambda-1}\frac{\theta[{\alpha \atop \beta}](z-w-(2s(\lambda)-1+\overline{\lambda(l-1)})P_-+\overline{\lambda(l-1)}P_+-(2\lambda-1)\Delta)}{\theta[{\alpha \atop \beta}](-(2s(\lambda)-1+\overline{\lambda(l-1)})P_-+\overline{\lambda(l-1)}P_+-(2\lambda-1)\Delta)}. \tag{3.8}$$

ecall that if $\lambda\notin Q$, then $\alpha_i,\beta_i\in[0,1]$ whereas if $\lambda=\frac{m}{n}$ then $\alpha_i,\beta_i\in\{0,\frac{1}{n},\ldots,\frac{n-1}{n}\}$. To ow that equations (3.6) and (3.8) coincide, we consider the propagator $S^{(l)}(z,w)$ in (3.8) λ - differential in z and expand it in the basis $f_j^{(\lambda)}(z)$

$$S^{(l)}(z,w)=\sum_j a^j(w)f_j^{(\lambda)}(z), \tag{3.9}$$

here

$$a^j(w)=\frac{1}{2\pi i}\oint_{C_\tau} S^{(l)}(z,w)f^j_{(1-\lambda)}(z), \tag{3.10}$$

i.e.

$$a^j(w) = \begin{cases} f^j_{(1-\lambda)}(w), & \text{if } j \leq Q(l,\lambda), \\ 0, & \text{if } j \geq Q(l,\lambda)+1, \end{cases} \qquad \tau_z > \tau_w;$$

(3.11

$$a^j(w) = \begin{cases} 0, & \text{if } j \leq Q(l,\lambda), \\ -f^j_{(1-\lambda)}(w), & \text{if } j \geq Q(l,\lambda)+1, \end{cases} \qquad \tau_w > \tau_z.$$

From eq.(3.10) it turns out that $S^{(l)}(z,w)$ can be seen as the generalization of the Szeg kernel to λ - and $(1-\lambda)$ - differentials in the w and z variables respectively.

From eqs.(3.5 - 6) it follows that

$$<0|\{b(z|l), c(w|l)\}|0> = \sum_j f_j^{(\lambda)}(z|l) f^j_{(1-\lambda)}(w|l). \qquad (3.1$$

The right - hand side is the delta function for λ - differentials, and being

$$\{b(z|l), c(w|l)\} = \Delta^{(l)}(z,w), \qquad (3.1$$

it follows that our previous definition, namely

$$<0|0> = 1, \qquad (3.1$$

is consistent. $|0>$ and its dual $<0|$ are the *natural* ket and bra vacua for the b - c system c a Riemann surface. These vacua are different from the vacua usually defined in the literatu [11]. For example, when $2\lambda(g-1)$ is an integer number, our procedure takes into account t existence of b's zero - modes in the correlations functions inserting them collectively in P When also the c's zero - modes are present they are inserted in P_+ [2]. To insert zero mod outside the points $P_\pm$, we expand the b field in terms of

$$g_j^{(\lambda)}(z|l) = \frac{E(z,P_+)^{j-s(\lambda)} \prod_{i=1}^{-2s(\lambda)+1} E(z,P_i)}{E(z,P_-)^{j-s(\lambda)+1}} \sigma^{2\lambda-1}(z)\cdot$$

$$\cdot\theta[^\alpha_\beta](z+(j-s(\lambda))P_+ - (j-s(\lambda)+1)P_- + \Sigma_{i=1}^{-2s(\lambda)+1} P_i + (1-2\lambda)\Delta), \qquad (3.1$$

where $j \in \mathbf{Z} + P(l,\lambda)$, $P(l,\lambda) = s(\lambda) + \lambda(l-1)$ and $-2s(\lambda)+1 \in \mathbf{Z}$. Its dual is

$$g^j_{(1-\lambda)}(z|l) = \frac{E(z,P_+)^{-j+s(\lambda)-1}}{E(z,P_-)^{-j+s(\lambda)} \prod_{i=1}^{-2s(\lambda)+1} E(z,P_i)} \sigma^{1-2\lambda}(z)\cdot$$

$$\cdot\theta[^{-\alpha}_{-\beta}](z-(j-s(\lambda)+1)P_+ + (j-s(\lambda))P_- - \Sigma_{i=1}^{-2s(\lambda)+1} P_i + (2\lambda-1)\Delta). \qquad (3.1$$

The propagator with the insertion of zero - modes in the points $P_1, ..., P_{-2s(\lambda)+1}$ is

$$S^{(l)}(z,w) = \frac{1}{E(z,w)} \left(\frac{E(z,P_-)}{E(w,P_-)}\right)^{-\overline{\lambda(l-1)}} \left(\frac{E(z,P_+)}{E(w,P_+)}\right)^{\overline{\lambda(l-1)}} \left(\prod_{i=1}^{-2s(\lambda)+1} \frac{E(z,P_i)}{E(w,P_i)}\right)\cdot$$

(3.1

$$\cdot\left(\frac{\sigma(z)}{\sigma(w)}\right)^{2\lambda-1} \frac{\theta[^\alpha_\beta](z-w-\overline{\lambda(l-1)}P_- + \overline{\lambda(l-1)}P_+ + \Sigma_{i=1}^{-2s(\lambda)+1} P_i - (2\lambda-1)\Delta)}{\theta[^\alpha_\beta](-\overline{\lambda(l-1)}P_- + \overline{\lambda(l-1)}P_+ + \Sigma_{i=1}^{-2s(\lambda)+1} P_i - (2\lambda-1)\Delta)}.$$

otice that to compute the correlation functions of fermion fields in the Ramond sector the nsertion of spin fields is automatically taken into account; actually any correlation function of λ - fields" connecting vacua in different sectors can be computed; they are the generalization f spin fields of the spin $\lambda \in \frac{\mathbb{Z}}{2}$ theory to $\lambda \in \mathbb{R}$. In general, the bosonized version of these elds are just the vertices of the minimal model theory [6].

4. KN algebra and KdV equation in higher genus

In this section we introduce the "covariantized KdV equation" and show that its second amiltonian structure is related to the KN algebra. This algebra is defined by

$$[e_i, e_j] = \sum_{s=-g_0}^{g_0} C_{ij}^s e_{i+j-s}, \qquad g_0 = \frac{3}{2}g, \tag{4.1}$$

here $e_j(z) \equiv f_j^{(-1)}(z)$ and

$$C_{ij}^s = \frac{1}{2\pi i}\oint_{C_\tau} f_{(2)}^{i+j-s}(z)[e_i(z), e_j(z)]. \tag{4.2}$$

he KN algebra admits central extension:

$$[e_i, e_j] = \sum_{s=-g_0}^{g_0} C_{ij}^s e_{i+j-s} + t\chi(e_i, e_j), \qquad [e_i, t] = 0, \tag{4.3}$$

here the cocycle is defined by

$$\chi(e_i, e_j) = \frac{1}{24\pi i}\oint_{C_\tau} \left(\frac{1}{2}(e_i'''e_j - e_j'''e_i) - \mathcal{R}(e_i'e_j - e_j'e_i)\right). \tag{4.4}$$

e projective connection $\mathcal{R}$ assures that the integrand is a well - defined 1 - form. If $\mathcal{R}$ has lar degree $mP_+ + nP_-$ with $m, n \leq 2$ then the cocycle satisfies the "locality" condition:

$$\chi(e_i, e_j) = 0 \quad for\ |i+j| > 3g. \tag{4.5}$$

turns out that this cocycle is unique up to trivial cocycles [1], in particular when $g = 0$ this ebra reduces to the Virasoro algebra. The Neveu - Schwarz and Ramond superalgebras in her genus have been constructed in ref.[4].

An alternative covariant expression for the integrand in eq.(4.4) can be obtained by mean an arbitrary vector field $\tilde{e}(z)$. The derivative of a λ - differential is covariantly well defined ly for $\lambda = 0$, therefore

$$\left(\left(\left(\frac{e_i}{\tilde{e}}\right)'\tilde{e}\right)'\tilde{e}\right)'\frac{e_j}{\tilde{e}} - \left(\left(\left(\frac{e_j}{\tilde{e}}\right)'\tilde{e}\right)'\tilde{e}\right)'\frac{e_i}{\tilde{e}}, \tag{4.6}$$

well - defined one - form, equal to twice the integrand in eq.(4.4) with

$$\mathcal{R} = \frac{\tilde{e}''}{\tilde{e}} - \frac{1}{2}\left(\frac{\tilde{e}'}{\tilde{e}}\right)^2. \tag{4.7}$$

If in a neighborhood of a point $P_i \in \Sigma$, $\tilde{e}(z) = z^a g(z)$, with $z(P_i) = 0$, the polar behaviour of the projective connection is

$$\mathcal{R} \sim \frac{a^2 - 2a}{2z^2} + a\frac{g'}{g}\frac{1}{z}. \tag{4.8}$$

Due to the poles in $P_i \neq P_\pm$, the cocycle $\chi(e_i, e_j)$ is τ - dependent; to get a τ - independent cocycle we define the "Baker - Akhiezer vector field"

$$\tilde{e}_H(z) = \frac{\sigma(P_+)\theta(P_+ - gP_- + \Delta)E(z, P_-)}{E(P_+, P_-)E(z, P_+)^{g-1}\sigma(z)\theta(z - gP_- + \Delta)} e^{-2\pi i \sum_1^\infty c_k \int_{P_0}^z \eta_k}, \tag{4.9}$$

where η_k is the normalized (i.e. $\oint_{a_i} \eta_k = 0,\ i = 1, ..., g$) second kind differential with poles of order $k+1$ in P_+ that in a local coordinate is

$$\eta_k(z) = \frac{1}{k!}\partial_w^k \partial_z lnE(z, w)dz_{|w=P_+}. \tag{4.10}$$

In order for $\tilde{e}_H(z)$ to be singlevalued, the constants c_k's must satisfy the equation

$$\Sigma_1^\infty c_k \oint_b \eta_k = 2\Delta - (g-1)(P_+ + P_-), \tag{4.11}$$

where $b \equiv (b_1, ..., b_g)$. Now we use $\tilde{e}_H$ in the framework of the second hamiltonian structure of the KdV equation formulated on higher genus Riemann surfaces (see also ref.[7]). The equation can be written in a covariant way using a procedure analogue to that used in eq.(4.7); in the following we briefly show how the procedure works. On the cylinder the KdV equation is

$$u_t = u''' + 6uu', \tag{4.12}$$

where u has conformal weight two and can be seen as the stress - energy tensor. Eq.(4.12) is closely related to the Virasoro algebra [8] whereas on higher genus Riemann surface it is related to the KN algebra. In order to show this relation let us start with the bi - hamiltonian form of eq.(4.12)

$$u_t = \mathcal{D}^{(i)}\frac{\delta\mathcal{H}^{(i)}}{\delta u}, \qquad i = 1, 2, \tag{4.13}$$

where

$$\mathcal{D}^{(1)} = \partial_x, \qquad \mathcal{H}^{(1)} = \frac{1}{2}\int dx(2u^3 - (u')^2), \tag{4.14}$$

$$\mathcal{D}^{(2)} = \frac{1}{2}(\partial_x^3 + 4u\partial_x + 2u'), \qquad \mathcal{H}^{(2)} = \int dx u^2. \tag{4.15}$$

Given two functionals $\mathcal{F}(x)$, $\mathcal{G}(y)$, their Poisson bracket is defined by

$$\{\mathcal{F}(x), \mathcal{G}(y)\}_{(i)} = \int dz \frac{\delta\mathcal{F}(x)}{\delta u(z)}\mathcal{D}^{(i)}(z)\frac{\delta\mathcal{G}(y)}{\delta u(z)}. \tag{4.}$$

Expanding $u(x)$ in a Fourier series, the Poisson bracket

$$\{u(x), u(y)\}_{(2)} = \mathcal{D}^{(2)}(x)\delta(x - y), \tag{4.}$$

yields the Virasoro algebra for the Fourier coefficients.

To write the KdV equation in a covariant way, we map the cylinder to the (punctured) phere and then insert the field $\tilde{e}_H$. The covariantized form of the operators $\mathcal{D}^{(2)}$ and $\mathcal{H}^{(2)}$ is

$$\mathcal{D}^{(2)}_{cov} = \frac{1}{2}\left(\tilde{e}_H^{-1}\partial_z \tilde{e}_H \partial_z \tilde{e}_H \partial_z + 4\tilde{e}_H u \partial_z + 2\left(u\tilde{e}_H^2\right)' \tilde{e}_H^{-1}\right)\tilde{e}_H^{-1}$$
$$= \frac{1}{2}\left(\tilde{e}_H^{-1}\partial_z \tilde{e}_H \partial_z \tilde{e}_H \partial_z \tilde{e}_H^{-1} + 4u\partial_z + 2u'\right),$$

$$\mathcal{H}^{(2)}_{cov} = \oint_{C_\tau} \tilde{e}_H^3 u^2, \tag{4.18}$$

here now u is a 2 - form on Σ. The KdV equation on Σ is

$$u_\tau = \mathcal{D}^{(2)}_{cov}\frac{\delta \mathcal{H}^{(2)}_{cov}}{\delta u} = \left(\left(\left(u\tilde{e}_H^2\right)' \tilde{e}_H\right)' \tilde{e}_H\right)' \tilde{e}_H^{-1} + 6u\left(u\tilde{e}_H^2\right)' \tilde{e}_H, \tag{4.19}$$

here

$$\frac{\delta u(x)}{\delta u(y)} = \Delta(x,y).$$

he covariant form of eq.(4.17) is

$$\{u(x),u(y)\}_{(2)} = \oint_{C_\tau} \Delta(x,z)\mathcal{D}^{(2)}_{cov}(z)\Delta(y,z) = \mathcal{D}^{(2)}_{cov}(x)\Delta(y,x). \tag{4.20}$$

xpanding the 2 - form $u(x)$ in the $f^j_{(2)}(x)$ basis and requiring that its coefficients satisfy the N algebra, we have

$$f^j_{(2)}(x)f^k_{(2)}(y)\{\tilde{L}_j,\tilde{L}_k\} = f^j_{(2)}(x)f^k_{(2)}(y)\left(C^s_{jk}\tilde{L}_{j+k-s} + \chi(e_j,e_k)\right). \tag{4.21}$$

nce the right hand sides of eqs.(4.20 - 21) coincide, it follows that the KdV equation on gher genus Riemann surfaces "generates" the KN algebra.

Let us now consider the Schrödinger equation on the cylinder

$$-\psi^{(\lambda)''} + u\psi^{(\lambda)} = -\lambda^2\psi^{(\lambda)}. \tag{4.22}$$

n important aspect of this equation is that it has symmetries generated by the charges H_n s explicit form will be given later). Indeed, it turns out that the parameter λ is unchanged der the shift

$$u(x) \to u(x) + \epsilon\{u(x),H_n\}_{(2)}. \tag{4.23}$$

e H_n's peculiarity is that they are in involution

$$\{H_n,H_m\}_{(2)} = 0. \tag{4.24}$$

The Schrödinger equation can be written in the following way

$$p^{(\lambda)2} + p^{(\lambda)'} = u + \lambda^2, \qquad p^{(\lambda)} = d\,ln\psi^{(\lambda)}, \tag{4.25}$$

at for large λ has the solutions

$$p^{(\lambda)} = \lambda + \sum_{n=1}^{\infty}\frac{\omega^{(n)}}{(2\lambda)^n}, \quad \tilde{p}^{(\lambda)} = -\lambda + \sum_{n=1}^{\infty}(-1)^n\frac{\omega^{(n)}}{(2\lambda)^n}, \tag{4.26}$$

where $\omega^{(n)}$ is determined by the recursion formula

$$\omega^{(1)} = u, \quad \omega^{(2)} = -u', \quad \omega^{(n)} = -\sum_{l=1}^{n-2} \omega^{(l)}\omega^{(n-1-l)} - \omega^{(n-1)'}. \tag{4.27}$$

It turns out that $\omega^{(2n)}$ is a total derivative, moreover eq.(4.24) is satisfied by

$$H_n = \int dx \omega^{(2n+1)}, \qquad n \geq 0. \tag{4.28}$$

To generalize this procedure to arbitrary genus, we write eq.(4.22) in a covariant way

$$-\left(\left(\psi^{(\lambda)'}\right)\bar{e}_H\right)'\bar{e}_H^{-1} + u\psi^{(\lambda)} = -\lambda^2\bar{e}_H^{-2}\psi^{(\lambda)}. \tag{4.29}$$

In higher genus, eq.(4.25) gets

$$p^{(\lambda)2} + p^{(\lambda)'} + p^{(\lambda)}\frac{\bar{e}'_H}{\bar{e}_H} = u + \lambda^2\bar{e}_H^{-2}. \tag{4.30}$$

The third equation in (4.27) is unchanged but the first two solutions are now

$$\omega^{(1)} = u\bar{e}_H, \qquad \omega^{(2)} = -\left(u\bar{e}_H^2\right)'. \tag{4.31}$$

Notice that starting with the covariantized expression of the Schrödinger equation we obta the same result for $\omega^{(n)}$ that covariantizing directly eq.(4.27). On Σ the H_n's are

$$H_n = \oint_{C_\tau} \omega^{(2n+1)}, \qquad n \geq 0. \tag{4.32}$$

The covariantization procedure shows that their Poisson bracket vanishes on Σ, indeed, sir the Poisson bracket vanishes on the cilynder, it is equal to the integral of a total derivati In our covariantization procedure a total derivative is seen as a one form therefore we ha to insert the "Baker - Akhiezer vector field" only on the right side of the derivative's symb that is any total derivative on the cylinder correspond to a total derivative on Σ.

Acknowledgements

I am grateful to L. Bonora for a careful reading of the manuscript and to L. Alvare Gaumè, F. Ferrari, R. Iengo, A.K. Raina, C. Reina for discussions and advices.

Appendix

In this appendix we recall some facts about theta functions theory [10]. The θ - funct with characteristic $[^\alpha_\beta]$ is defined by

$$\theta[^\alpha_\beta](z) = \sum_{n\in Z^g} e^{\pi i(n+\alpha)\Omega(n+\alpha)+2\pi i(n+\alpha)(z+\beta)}$$

$$= e^{\pi i\alpha\Omega\alpha+2\pi i\alpha(z+\beta)}\theta(z+\beta+\Omega\alpha), \tag{A}$$

$$\theta(z) = \theta[^0_0](z), \qquad z \in C^g, \quad \alpha, \beta \in R^g,$$

where $\Omega_{i,j} \equiv \oint_{b_i} \omega_j$, $\Omega_{i,j} = \Omega_{j,i}$, $Im(\Omega) > 0$. The holomorphic differentials ω_i, $i = 1, ..., g$ are normalized in such a way that $\oint_{a_i} \omega_j = \delta^i_j$, a_i, b_i being the homology cycles basis.

When $\alpha_i, \beta_i \in \{0, \frac{1}{2}\}$, the θ - function is even or odd depending on the parity of $4\alpha\beta$. The θ - function is multivalued under a lattice shift in the z - variable:

$$\theta[^\alpha_\beta](z + n + \Omega m) = e^{-\pi i m \Omega m - 2\pi i m z + 2\pi i(\alpha n - \beta m)} \theta[^\alpha_\beta](z). \tag{A.2}$$

Riemann vanishing theorem.

The function

$$f(z) = \theta(I(z) - \Sigma^g_{i=1} I(P_i) + I(\Delta)), \qquad z, P_i \in \Sigma \tag{A.3}$$

either vanishes identically or it has g simple zeroes in $z = P_1, ..., P_g$.

Δ is the Riemann divisor class defined by

$$I_k(\Delta) = \frac{1 - \Omega_{k,k}}{2} + \sum_{j \neq k} \oint_{a_j} \omega_j(z) I_k(z), \tag{A.4}$$

where

$$I_k(z) = \int_{P_0}^{z} \omega_k, \qquad P_0, z \in \Sigma, \tag{A.5}$$

is the Jacobi map (P_0 is an arbitrary reference point) and

$$I(D = \Sigma^n_{i=1} m_i P_i) \equiv \Sigma^n_{i=1} m_i I(P_i), \qquad m_i \in R.$$

Another useful theorem, due to Riemann, states that

$$2\Delta = [K] \tag{A.6}$$

where K is the canonical line bundle and $[K]$ denotes the associated divisor class. We recall that two divisors D_1, D_2 belong to the same divisor class $[D]$ if $D_1 - D_2$ is equal to the divisor of a meromorphic function.

Abel theorem.

Let D be a divisor on Σ. Then

$$I(D) = I([D]) \ mod. \Gamma \equiv \{v \in C^g \,|\, v = n + \Omega m, \ \ n, m \in Z^g\}. \tag{A.7}$$

The prime form is defined by

$$E(z, w) = \frac{\theta[^\alpha_\beta](I(z) - I(w))}{h(z) h(w)} = -E(w, z), \qquad z, w \in \Sigma, \tag{A.8}$$

is a holomorphic (multivalued) $(-\frac{1}{2}, -\frac{1}{2})$ - differential with a simple zero in $z = w$:

$$E(z, w) \sim z - w, \ as \ z \to w. \tag{A.9}$$

$h(z)$ is the square root of $\Sigma_{i=1}^{g}\omega_i(z)\partial_{u_i}\theta[{\alpha \atop \beta}](u)|_{u_i=0}$, it is the holomorphic $\frac{1}{2}$ - differential wit non singular (i.e. $\partial_{u_i}\theta[{\alpha \atop \beta}](u)|_{u_i=0} \neq 0$) odd spin structure $[{\alpha \atop \beta}]$. Notice that $E(z,w)$ does n depend on the particular choice of $[{\alpha \atop \beta}]$. The prime form has the following multivaluedne around the b's homology cycles:

$$E(z+na+mb,w) = e^{-\pi i m\Omega m - 2\pi i m(I(z)-I(w))}E(z,w). \qquad (A.1$$

The σ - differential is defined by

$$\theta(z - I(P_1 + ... + P_g) + \Delta) = s(P_1,...,P_g)\sigma(z)E(z,P_1)...E(z,P_g), \qquad (A.1$$

where $s(P_1,...,P_g)$ is a holomorphic section of a line bundle of degree $g-1$ in each variab

References

[1] I.M. Krichever and S.P. Novikov, Funk. Anal. i Pril., 21 No.2 (1987) 46 and No.4 (198 47.

[2] L. Bonora, A. Lugo, M. Matone and J. Russo, "A global operator formalism on high genus Riemann surfaces. b - c systems", preprint SISSA 67/88/EP, to appear in Com Math. Phys.

[3] L. Bonora, M. Matone and M. Rinaldi, "Relation between representations of KN a Virasoro algebras", preprint SISSA 119/88/EP, to appear in Phys. Lett. B.

[4] L. Bonora, M. Bregola, P. Cotta - Ramusino and M. Martellini, Phys. Lett. B205 (198 53; L. Bonora, M. Martellini, M. Rinaldi and J. Russo, Phys. Lett. B206 (1988) 444.

[5] L. Bonora, M. Rinaldi, J. Russo and K. Wu, Phys. Lett. B208 (1988) 440.

[6] L. Bonora, M. Matone and K. Wu, in preparation.

[7] L. Bonora, M. Matone, "KdV equation on higher genus Riemann surfaces", to appe

[8] J. L. Gervais and A. Neveu, Nucl. Phys. B209 (1982) 125; J. L. Gervais, Phys. Le B160 (1985) 277, 279; P. Mathieu, Phys. Lett. B208 (1988) 101.

[9] H. Farkas and I. Kra, "Riemann surfaces". Springer, 1980.

[10] J. Fay, "Theta Functions on Riemann Surfaces", Lectures Notes in Mathematics 3 Springer - Verlag (1973); D. Munford, "Tata Lectures on Theta", Vol. I, II. Birkhaus Boston (1983).

[11] L. Alvarez - Gaumè, C. Gomez and C. Reina, "New methods in string theory", prep CERN - TH 4775/87; L. Alvarez - Gaumè, C. Gomez, G. Moore and C. Vafa, Nu Phys. B303 (1988) 455, and references therein.

Supermoduli and Superstrings *

Gregorio Falqui and Cesare Reina

S.I.S.S.A

(International School for Advanced Studies)

Strada Costiera 11, 34014 - TRIESTE (Italy)

ABSTRACT

e recall some deformation theory of susy–curves and construct the local model of their (compactified) oduli 'spaces'. We also construct universal deformations "concentrated" at isolated points, which e the mathematical counterparts of the usual choices done in the physical literature. We argue that ese cannot give a projected "atlas" for supermoduli spaces.

1. Introduction

Super–Riemann surfaces have been introduced by Friedan [F] in view of applications to superstring eory and superconformal field theories. Although there may be other possible set ups, the simplest athematical way of understanding such objects comes from the theory of graded manifolds in the ase of Kostant and Leites (see e.g. [L]). Besides matching with physics, this approach allows a direct tension of the standard methods of algebraic geometry to the $\mathbf{Z}_2$-graded commutative case [M][D]. deed, (compact) super Riemann surfaces are algebraic objects – the so-called susy-curves.

A good deal of work has been already done on the geometry of susy-curves, generalizing most of e results which hold true for ordinary curves. Among the open problems, a topic which has recently ined a central position in the physical literature has to do with 'moduli' of susy-curves, because of eir relevance in defining amplitudes in superstring theory (see [DP] and references quoted therein). ese so-called supermoduli spaces have been studied also in the mathematical literature [LR][W], but e basic question asked by physicists, has still no definite answer - namely it is not known whether one n define a sensible integration theory on these complex superspaces. A possible way out [MT] comes m Rothstein's results for the real case [R2]. These can be applied almost verbatim to integration er complex supermoduli spaces, provided they are 'projected'.

Besides physical applications, the study of the structure of supermoduli spaces has its own math-atical interest, and this paper is intended to give a contribution in this direction. In sect.2 we recall ne deformation theory of susy-curves, which will give us both the reduced spaces and the local mod- of supermoduli spaces. These reduced spaces are the same as the moduli spaces of θ-characteristics, ich will be described in sect.3 together with their compactifications recently found by Cornalba [C]. re we also compute the natural extension of the Mumford formula [M] to the boundary of moduli ces. Finally, in sect.4 we tackle the problem of the global structure of supermoduli spaces. The in idea we pursue here is to construct universal deformations of susy-curves, which depend 'linearly' odd-modular parameter, by mimicking Schiffer deformations of ordinary curves. In this way one ds an explicit representative for the obstruction class to 'projectedness'. This is easily shown [FR] be trivial at genus $g \leq 2$, getting another proof of known results. The much harder question about triviality of this class for $g > 2$ is still under investigation.

As a world of caution, we will work as the supermoduli functor were representable, that is a ermoduli spaces exist as supermanifolds. Instead, what we are actually describing is the analogue the moduli stack for ordinary curves. We will leave to future research the construction of its full rse structure.

Work partially supported by the national project "Geometria e Fisica" M.P.I

2. Susy-curves

It is nowadays well assessed [D,GN,LR,M] that the geometrical structure underlying two dime sional superconformal supergravity can be encoded in the

Definition 2.1 . A family of (smooth) susy-curves X parameterized by a complex superspace or, for the sake of brevity, a susy-curves X over B, is a proper surjective map $\pi: X \longrightarrow B$ of compl superspaces having $1|1$-dimensional fibres, together with a $0|1$-dimensional distribution $\mathcal{D}_\pi$ in t relative tangent sheaf $\mathcal{T}_\pi X$ such that the supercommutator *mod* $\mathcal{D}_\pi$, $[\ ,\ ]_{\mathcal{D}} : \mathcal{D}_\pi^{\otimes^2} \longrightarrow \mathcal{T}_\pi X/\mathcal{D}_\pi$ is isomorphism.

When B reduces to a point $\{*\}$, one will speak of 'isolated' or 'single' susy-curves. The li between the above definition and the usual (see, e.g. [F]) field-theoretic approach to superconform models is as follows. A relative coordinate system on the susy-curve $X \xrightarrow{\pi} B$ given by coordinate cha $\{(U_\alpha, z_\alpha, \theta_\alpha, b)\}$ is called canonical if the local generator for $\mathcal{D}_\pi$ is expressed as $D_\alpha = \frac{\partial}{\partial \theta_\alpha} + \theta_\alpha \frac{\partial}{\partial z_\alpha}$ Then $\mathcal{D}^{\otimes 2}$ is locally generated by $\frac{\partial}{\partial z_\alpha}$. As shown by [LR], any susy-curve admits a canonical atlas

Remark 2.2 . For the case of a single susy-curve, a short computation shows that requiring t existence of such a distribution $\mathcal{D}_\pi$ is tantamount to imposing the structure sheaf of X to be of t form $\mathcal{O} \oplus \Pi\mathcal{L}$ where $\mathcal{L}$ is a (relative) θ-characterisitcs and Π is the so-called parity-changing funct whose effect is to make sections of $\mathcal{L}$ anticommute.

Besides being mathematically natural, the need for families of susy-curves in physical applicatio follows from the fact that world-sheet supersymmetry requires the presence of a gravitino field o given single susy-curve C. One can fix local superconformal gauges [Ho], which amount to choosi local complex coordinates, a local holomorphic trivialization of $\mathcal{L}$ and to identifying a chiral piece of gravitino field with a section χ of $A^{0,1}(C_{red}, \mathcal{L}^{-1})$, i.e. with a smooth antiholomorphic one form w values in $\mathcal{L}^{-1}$ or, passing to Čech cohomology, with a Čech 1–cocycle $\epsilon_{\alpha\beta}$ with values in $\mathcal{L}^{-1}$. Not that the action of supersymmetry has no effect on the $\epsilon_{\alpha\beta}$'s, while we have a local symmetry genera by holomorphic sections η_α of $\mathcal{L}^{-1}$ acting via Čech coboundaries, i.e. as $\epsilon_{\alpha\beta} \rightarrow \epsilon_{\alpha\beta} + \eta_\alpha - \eta_\beta$. In ot words, we can benefit of the isomorphism $H^{0,1}_{\bar\partial}(C_{red}, \mathcal{L}^{-1}) = H^1(C_{red}, \mathcal{L}^{-1})$ to represent gravit fields χ (up to supersymmetries) via Čech cocycles $\epsilon_{\alpha\beta}$ (up to coboundaries).

The datum of $[\epsilon]$ can be encoded in an extension of the structure sheaf of C as follows. Forgett about parity, consider $H^1(C_{red}, \mathcal{L}^{-1})$ as a constant sheaf with group $\mathbf{C}^{2g-2}$ on C_{red}. If $\epsilon_{\alpha\beta}$ i representative of $[\epsilon]$ and $\epsilon^i_{\alpha\beta}$ represent a basis $[\epsilon^i]$ for $H^1(C_{red}, \mathcal{L}^{-1})$, we set $\epsilon_{\alpha\beta}(\zeta_i) = \epsilon^i_{\alpha\beta}\zeta_i$, (sum c $i = 1, \ldots, 2g-2$) and construct an extension $\mathcal{F}$ of $\mathcal{L}$ by $\mathbf{C}^{2g-2}$ by stating that $\mathcal{F}$ is the sheaf of secti of a rank $2g-1$ vector bundle locally generated by θ_α, ζ_i with transition functions *

$$\begin{pmatrix} \theta_\alpha \\ \zeta_i \end{pmatrix} = \begin{pmatrix} \pm\sqrt{f'_{\alpha\beta}} & \epsilon^i_{\alpha\beta} \\ 0 & 1 \end{pmatrix} = \begin{pmatrix} \theta_\beta \\ \zeta_i \end{pmatrix}$$

Notice that $\mathcal{F}$ is independent (up to isomorphisms) both of the basis $[\epsilon^i]$ and of its representatives. supermanifold $(C_{red}, \wedge\mathcal{F})$ is not yet a susy-curve, but we can cook out of the same data a deforma $\mathcal{A}$ of $\wedge\mathcal{F}$ making $(C_{red}, \mathcal{A})$ a susy-curve. It is enough to find a superconformal coordinate patc $z_\alpha = z_\alpha(z_\beta, \theta_\beta, \zeta_i)$, $\theta_\alpha = \theta_\alpha(z_\beta, \theta_\beta, \zeta_i)$ which reproduces the transition functions above for $\mathcal{F}$ $z_\alpha = f_{\alpha\beta}(z_\beta) \bmod \mathcal{N}^2$. (here $\mathcal{N}$ is the nilpotent ideal locally generated by θ_α, ζ_i). The "minin answer is

$$\begin{cases} z_\alpha = f_{\alpha\beta}(z_\beta) + \theta_\beta \sqrt{f'_{\alpha\beta}(z_\beta)}\,\epsilon_{\alpha\beta}(z_\beta, \zeta_i) \\ \theta_\alpha = \sqrt{f'_{\alpha\beta}(z_\beta) + \epsilon_{\alpha\beta}(z_\beta)\epsilon'_{\alpha\beta}(z_\beta)} \cdot \theta_\beta + \epsilon_{\alpha\beta}(z_\beta, \zeta_i) \end{cases}$$

By minimal here we mean that it depends only on the data already encoded in $\mathcal{F}$ at the lo order compatible with superconformal structures. Unfortunately, we see that, in spite the local m

* Hereinafter $f'_{\alpha\beta}$ means $\partial f_{\alpha\beta}/\partial z_\beta$ The sign ambiguity refers to the choice of a θ-characteristic C and will be left implicit in the following.

$\mathcal{F}$ was independent of choices, $\mathcal{A}$ is not. In particular it is not independent of the choice of the presentatives $\epsilon^i_{\alpha\beta}$ because of the non-linear therm $\epsilon\epsilon'$ entering the transition functions. In any case, $(\mathcal{C}_{red}, \mathcal{A})$ gives us an example of a non trivial susy-curve encoding informations about gravitino fields.

A first step in the construction of supermoduli spaces is to study some deformation theory of susy-urves. We refer to Waintrob [W] for the general set up of deformation theory of complex superspaces. et us simply remark here that, as susy curves are more than generic 1|1-dimensional superspaces, eir deformations must be defined as follows;

efinition 2.3 . A deformation of a single susy-curve C over (the germ of a pointed) complex perspace (B, b_0) at $b_0 \in B$ is a family $\pi: X \longrightarrow B$ of susy-curves over B together with a fixed omorphism $i : C \to \pi^{-1}(b_0)$ between C and the special fibre over b_0.

This makes sense because each fibre $\pi^{-1}(b), b \in B$, is itself a single susy curve with the subsheaf induced by $\mathcal{D}_\pi$. Notice that, because of remark 2.2, an isomorphism of single susy-curves may be ought of as induced by an isomorphism of the underlying θ-characteristics. Notice also that fixing e isomorphism $i : C \to \pi^{-1}(b_0)$ is vital as in the ordinary case, since it allows the study of the action the automorphism group of C on the base space B of its deformations.

Let us recall that, given a deformation $\pi : X \to B$ of a susy-curve, we have two natural subsheaves the tangent sheaf $\mathcal{T}X$. Along with the relative tangent sheaf $\mathcal{T}_\pi X = \ker \pi_*$ there is the sheaf $\mathcal{T}^{\mathcal{D}}X$ of rivations which commute with sections of $\mathcal{D}_\pi$. A basic role is played by the sheaf $\mathcal{T}^{\mathcal{D}}_\pi =: \mathcal{T}_\pi X \cap \mathcal{T}^{\mathcal{D}}X$ infinitesimal automorphisms of X. One can show [LR] that there is an isomorphism $\mathcal{T}^{\mathcal{D}}_\pi \simeq \mathcal{D}^{\otimes 2}_\pi$. early enough, the sequence

$$0 \to \mathcal{T}^{\mathcal{D}}_\pi X \to \mathcal{T}^{\mathcal{D}}X \to \mathcal{T}B \to 0$$

exact.

Mimicking what happens in standard deformation theory, one first studies infinitesimal de-mations. To this purpose, one introduces the super-commutative ring of super-dual numbers $= \mathbf{C}[t,\zeta]/(t^2, t\zeta)$, where $(t,\zeta) \in \mathbf{C}^{1|1}$, $\mathbf{C}[t,\zeta]$ is the polynomial ring and $(t^2, t\zeta)$ is the ideal gener-ed by t^2 and $t\zeta$. Associated to this ring there is a superspace $S = (\{*\}, \mathcal{O}_S)$, which embodies the a of a super-tangent vector. A deformation of C over S will be called an infinitesimal deformation.

Given a complex superspace (B, b_0), the tangent space $T_{b_0}B$ at b_0 is isomorphic to the linear perspace $Mor(S,B) = \{f : S \to B \mid f(*) = b_0\}$ of superspace morphisms. Now, given a deformation $\to B$ of C, we can think of a tangent vector $[f]$ in $T_{b_0}B$ as a map $f \in Mor(S,B)$ and the pull-back formation $f^*X \to S$ is a first order deformation of C. The Kodaira-Spencer class of this tangent ctor is obtained by considering the exact sheaf sequence

$$0 \to f^*\mathcal{T}^{\mathcal{D}}_\pi X \longrightarrow f^*\mathcal{T}^{\mathcal{D}}X \longrightarrow f^*\mathcal{T}B \to 0$$

king the coboundary map one has

$$\widehat{KS}_f : H^0(f^*\mathcal{T}B) \equiv \{[f] \in T_{b_0}B\} \to H^1(f^*\mathcal{T}^{\mathcal{D}}_\pi X) \equiv H^1(C, \mathcal{D}^{\otimes 2}).$$

tting f vary we get the Kodaira-Spencer homomorphism

$$\widehat{KS} : T_{b_0}B \to H^1(C, \mathcal{D}^{\otimes 2}).$$

Brun and Rothstein [LR] proved that a family of susy-curves for which KS is an isomorphism is odular'. On the other hand there is merit in considering deformations, because as in the ordinary e, one can prove the following

eorem 2.4 . A deformation of a susy-curve $C \overset{i}{\hookrightarrow} X \overset{\pi}{\longrightarrow} \Delta$ for which $\widehat{KS}$ is an isomorphism is versal.

oof. Promoting a modular family to be a deformation by adding the datum of the isomorphism $\to \pi^{-1}(0)$, helps in killing the possible $\mathbf{Z}_2$ ambiguities envisaged by [LR] (see prop. 2.6 and 2.7). In fact a deformation of $(C_{red}, \wedge\mathcal{L})$ over a reduced base is the same as a deformation of a θ-characteristi . So if $C \overset{j}{\hookrightarrow} Y \overset{\pi'}{\longrightarrow} \Delta'$ is any deformation of $(C, \wedge\mathcal{L})$ there exists a unique $f_{red}\colon \Delta'_{red} \longrightarrow \Delta_{red}$ such th the diagram

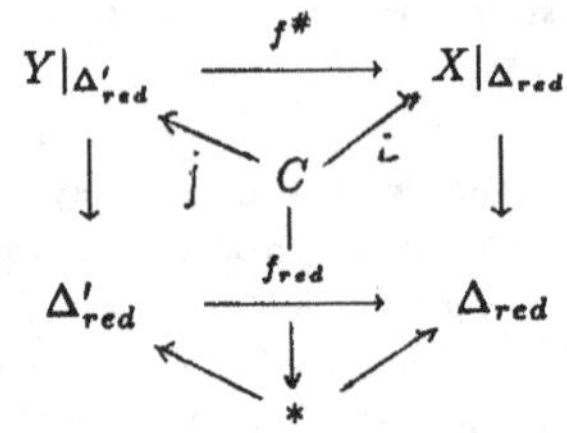

commutes. Notice that $f^\#$ is uniquely fixed by i and j. Hence proposition 2.6 of [LR] tells us th there is a unique extension $f\colon \Delta' \longrightarrow \Delta$ of f_{red} and an isomorphism ψ such that the diagram

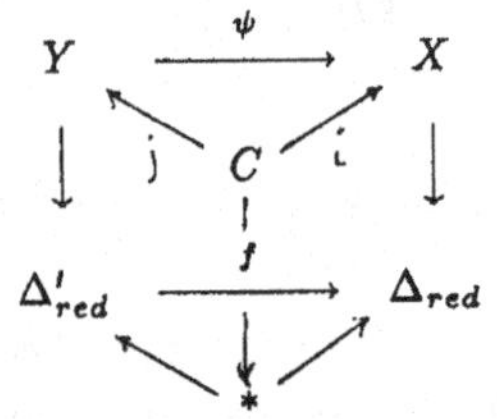

commutes as well. The only possible ambiguity concerns now the uniqueness of the isomorphism ψ ψ_1, ψ_2 were two such isomorphisms, then $\psi_1 \circ \psi_2^{-1}$ is either the identity or the canonical automorphi of Y (prop. 2.7 of [LR]). But the commutativity of the latter diagram fixes it as $\psi_1 \circ \psi_2^{-1} = \mathrm{id}_Y$.

As a consequence we have that the group of automorphisms of a susy-curve naturally acts on base space of its universal deformations - a fact which is vital in constructing the coarse structure supermoduli spaces.

To classify infinitesimal deformations of susy-curves in the spirit of the "original" Koda Spencer approach one can proceed as follows, regarding a susy-curve as built by patching toget 1|1-dimensional superdomains by means of superconformal transformations, and singling out 'infinit imal' moduli as non-trivial parameters in the transition functions. Namely, consider a canonical at $\{U_\alpha, z_\alpha, \theta_\alpha,\}$ for C with clutching functions

$$\begin{cases} z_\alpha = f_{\alpha\beta}(z_\beta) \\ \theta_\alpha = \sqrt{f'_{\alpha\beta}}\theta_\beta \end{cases}$$

They obviously satisfy the cocycle condition $f_{\alpha\beta}(f_{\beta\gamma}(z_\gamma)) = f_{\alpha\gamma}(z_\gamma)$ on $U_\alpha \cap U_\beta \cap U_\gamma$.

We can cover a first order deformation $\pi : X \to S$ of C glueing the $U_\alpha \times S$ via the identificat

$$z_\alpha = f_{\alpha\beta}(z_\beta) + t b_{\alpha\beta}(z_\beta) + \theta_\beta \zeta g_{\alpha\beta}(z_\beta) F_{\alpha\beta}(z_\beta)$$

$$\theta_\alpha = F_{\alpha\beta}\theta_\beta + \zeta g_{\alpha\beta}$$

where $F_{\alpha\beta} = \sqrt{f'_{\alpha\beta} + t b'_{\alpha\beta}}$, so that the clutching functions are superconformal for any t, ζ. The coc condition for these transformation rules reduce to the cocycle condition for the $f_{\alpha\beta}$'s as before, p

$$b_{\alpha\beta} + f'_{\alpha\beta}\, b_{\beta\gamma} = b_{\alpha\gamma}$$

$$g_{\alpha\beta}\theta_\alpha + f'_{\alpha\beta}\, g_{\beta\gamma}\theta_\beta = g_{\alpha\gamma}\theta_\gamma$$

aking the tensor product by $\partial/\partial z_\alpha$, one sees that the one cochains

$$v^0_{\alpha\beta} = \{b_{\alpha\beta}\partial/\partial z_\alpha\}$$

$$v^1_{\alpha\beta} = \{g_{\alpha\beta}\theta_\alpha \otimes \partial/\partial z_\alpha\}$$

e actually cocycles. They define a class in

$$H^1(C_{red}, \omega^{-1}) \oplus \Pi H^1(C_{red}, \mathcal{L}^{-1}) = H^1(C, \mathcal{D}^{\otimes 2}) \subset H^1(C, TC)$$

lled the Kodaira-Spencer class of the first order deformation* $X \xrightarrow{\pi} S$.

A similar computation, considering local superconformal reparametrizations with local odd pa-meters λ_α, shows that they leave the cocycle v_0 invariant and send v_1 into $\tilde{v}^1_{\alpha\beta} = v^1_{\alpha\beta} + (\lambda_\alpha - \sqrt{f'_{\alpha\beta}} \cdot$ $)\theta_\alpha \otimes \frac{\partial}{\partial z_\alpha}$ which leads to the

heorem 2.5 . ([FMS],[M],[LR]). The set of equivalence classes of first order deformations of a sy-curve C is a linear complex superspace with dimension $3g-3|2g-2$.

roof. It is enough to compute the dimensions of $H^1(C_{red}, \omega^{-1})$ and $H^1(C_{red}, \mathcal{L}^{-1})$ by means of iemann-Roch theorem.

Since C is split, $\mathcal{D}^{\otimes 2} \simeq TC_{red} \oplus \Pi\mathcal{L}^{-1}$, and $H^1(C, \mathcal{D}^{\otimes 2})$ naturally splits into even and odd bspaces and we can speak about even and odd Kodaira-Spencer homomorphisms KS_0 and KS_1, composing $\widehat{KS}$ with the projections of $H^1(C, \mathcal{D}^{\otimes 2}) = H^1(C_{red}, \omega^{-1}) \oplus \Pi H^1(C_{red}, \mathcal{L}^{-1})$ onto the st and second summand. It follows that, if B is a purely even superspace (i.e. an ordinary complex ace, see [W]), $KS_1 = 0$ and $KS_0; T_{b_0}B \to H^1(C_{red}, \omega^{-1})$ is the ordinary Kodaira-Spencer map. sing the natural map $i: B_{red} \to B$, we get that a deformation $X \to B$ is versal on a purely even B, and only if the induced deformation $i^*X \to B_{red}$ is. As we need the datum of a θ-characteristics on $_{red}$, the deformation $X \to B_{red}$ has to be be considered as a deformation of a θ-characteristics. We ve therefore the following

roposition 2.6 . Even-versal deformations of a SUSY-curve exist and are in 1-1 correspondence th pull-backs under maps $f: B \to B_{red}$ of versal deformations over B_{red} of the underlying θ-aracteristic.

This proposition tells us that the reduced space of moduli of susy-curves can be given in terms isomorphism classes of pairs $(C_{red}, \mathcal{L})$. As we shall see in the next section, this space S_g comes uipped with the "universal curve" $\pi: \mathcal{C} \to S_g$ together with the "universal (dual) θ-characteristics" $^{-1}_\pi \to \mathcal{C}$ on $\mathcal{C}$. From the construction above it is then clear that the first infinitesimal neighbourhood supermoduli space is the sheaf $R^1\pi_*(\mathcal{L}_\pi^{-1})$ on S_g. Accordingly the local model for supermoduli aces is given by the split supermanifold $(S_g, \wedge R^1\pi_*\mathcal{L}_\pi^{-1})$. We will discuss in sect. 4 how much tual supermoduli spaces may differ from being split.

3. Moduli space of θ-characteristics

As already remarked, from a susy-curve we inherit a curve C_{red} (which will now be simply denoted C, since supercommutative objects will never appear in this section and no confusion can rise) plus θ-characteristics $\mathcal{L}$ on it. When C is smooth, $\mathcal{L}$ is a square root of the canonical sheaf ω of C, i.e. $= \omega$. Recall that such an $\mathcal{L}$ is called even or odd according to the parity of $dim\ H^0(C, \mathcal{L})$. There e $2^{g-1}(2^g+1)$ even and $2^{g-1}(2^g-1)$ odd θ-characteristics, adding to a total of 2^{2g}. A survey of ese and other relevant properties may be found in [ACGH] and references quoted therein.

A deformation of a (smooth) θ-characteristics $(C, \mathcal{L})$ is a deformation $\pi: X \to B$, together with invertible locally free sheaf $\mathcal{L}_\pi$ on X such that $\mathcal{L}_\pi^2$ is isomorphic to the relative canonical sheaf ω_π d the isomorphism $i: C \to \pi^{-1}(b_0)$ induces an isomorphism of $\mathcal{L}$ and $i^*\mathcal{L}_\pi$.

---

* Here we obviously assume that C is smooth. Deformation theory of SUSY-curves with nodes quires the handling of θ -characteristics in the singular case, and will be dealt with elsewhere.

Let us now see what happens to θ-characteristics, when a curve C degenerates to a node cur during a deformation. We will start with the simplest example.

Example 3.1 . Let us consider the family of elliptic curves parameterized by a small disk $\Delta \subset \mathbf{C}$ follows. Set $\tau = ln(b)/2\pi i$, $b \in \Delta$, and consider the lattice $\Lambda_\tau \subset \mathbf{C}$ generated by 1 and τ. This ac holomorphically on $\Delta \times \mathbf{C}$ by translations on the second factor. The quotient $X = \Delta \times_{\Lambda_\tau} \mathbf{C}$ is a fami of tori degenerating to a single-node curve for $b = 0$. At genus one all θ-characteristics have degree and one of them is isomorphic to the structure sheaf $\mathcal{O}$. So the other three naturally corresponds points of order two on the Jacobian, which in turn coincides with the torus itself. So, on $\Delta - \{0\}$ get the following sections of $J = X \to \Delta$

$$\sigma_1 = 0; \qquad \sigma_2 = 1/2$$

$$\sigma_3 = \tau/2; \qquad \sigma_4 = \tau/2 + 1/2$$

$(mod\, \Lambda_\tau)$, where $\tau = \tau(b)$ as above. We can now clearly see three phenomena. First of all, we ha monodromy in the covering, because a rotation around $b = 0$ exchanges the two sections σ_3 and Second, these two sections are 'asymptotic' for $b \to 0$ ($|\tau| \to \infty$), meaning that there is branchi in the covering (recall that the Jacobian of a torus with one node can be compactified getting ag the same torus; being asymptotic here means that the two section above go to the node in the limi Finally, this limit point cannot be interpreted any more as an invertible sheaf, but corresponds t more general coherent sheaf.

If we abstract from the peculiarities of genus 1, the picture we get from this example is gener In particular, the three phenomena mentioned above, i.e. monodromy, branching and the appeara of more general sheaves than sheaves of sections of line bundles reproduce themselves at all gene For instance, such sheaves occur in the compactification of the moduli of θ-characteristics recen constructed by Deligne [D]. Another way of getting compactified moduli spaces has been found Cornalba [C]; this involves a slightly wider class of singular curves to be added as the boundary moduli spaces and has the advantage that θ-characteristics have invertible sheaves as 'limit' on t boundary. As we feel this property quite natural for physical applications, in the sequel we will disc some of its features.

In a certain sense, the whole construction stems from the observation that the appearance monodromy and of non locally free sheaves are somewhat related. In fact, let $\pi : X \to \Delta$ be a parameter deformation of a stable curve C with smooth fibres $\pi_{-1}(t)$, ($t \in \Delta - \{0\}$) and assume simplicity that C itself has a single node. In other words, $t \in \Delta$ is a local coordinate transversa some component δ_i of the boundary of the ordinary moduli space $\overline{M}_g$. The local equation of X n the node of the central fibre can be written as $xy = t$. It follows that, in spite the central fibr singular, the (2-complex dimensional) surface X is smooth. Next, assume a family of θ-characteris $\mathcal{L}'_\pi$ is given on $X - \pi^{-1}(0)$, and ask whether it can be extended to the whole of X. We can get ri monodromy, if present, by double covering the base of X, i.e. by setting $t = f(q) = q^2$ and pulli back X to get a deformation $Y = f^*X$ over another disk Q. The local equation for Y now re $xy = q^2$, which clearly shows that Y is singular at the node on the central fibre. So $f^*\mathcal{L}'_\pi$ cannot extended as an invertible sheaf. To get such an extension, one first smooths out Y by blowing up singular point. The family $Z \to Q$ got in this way is the same as Y off the singular point, while latter has been substituted by an entire line E (a copy of the Riemann sphere), called the excepti line. Thus the central fibre is now a semistable curve C_0. Its normalization has components C' E given respectively by the normalization of C and by the exceptional line E. On C_0, E and intersect in two points p_1, p_2 given by the preimages on C' of the node on C. If a, b $(a = 1/b)$ local coordinates on E, the blow up is given by $ax = q$, $by = q$ which shows the presence of nodes at $q = 0$. In spite that C has been replaced by an even more singular curve C_0, now Z is sm and $\mathcal{L}'_\pi$ can be extended to an invertible sheaf $\mathcal{L}_\pi$ on the whole of Z; we denote by $\mathcal{L}_0$ the shea get in this way on the central fibre.

Clearly enough, such an extension $\mathcal{L}_\pi$ is not unique, because by tensoring with any sheaf of the rm $\mathcal{O}(nE)$ one gets another extension. The basic fact which matters for us is that one can choose e extension $\mathcal{L}_\pi$ so that the restriction $\mathcal{L}_0|_E$ of $\mathcal{L}_\pi$ to E is isomorphic to $\mathcal{O}(n)$ with n either 0 or 1. see why this is so, assume that $\mathcal{L}_0|_E$ was $\mathcal{O}(s)$, then $\mathcal{L}_\pi(nE)$ restricts to $\mathcal{L}_0|_E(-np_1 - np_2)$ which then isomorphic to $\mathcal{O}(s-2n)$. Therefore, by suitably choosing n the degree of $\mathcal{L}_0|_E$ can be adjusted be either 0 or 1.

Let's now see the relations between $\mathcal{L}_\pi$ and θ-characterisitcs. For $q \neq 0$, $\mathcal{L}_q^2 = \omega_q$, where as usual e subscript q indicates the restriction to the fibre of Z over $q \in Q$. So $deg\mathcal{L}_q = g-1$ and the same true for $\mathcal{L}_0$. We have thus two cases [C]

roposition 3.2 . Let ω_0 be the dualizing sheaf of C_0.

if $deg\mathcal{L}_0|_E = 0$ (and then $deg\mathcal{L}_0|_{C'} = g-1$) we have that $\mathcal{L}_0^2 = \omega_0$.

if $deg\mathcal{L}g_0|_E = 1$ (and then $deg\mathcal{L}_0|_{C'} = g-2$) we have that $\mathcal{L}_0^2(E) = \omega_0$

roof. We first recall the intersection properties of the divisors C' and E on the surface Z. Since $' + E$ is homologically equivalent to a generic fibre Z_q which does not intersect either C' or E, we ve $0 = E.(C'+E) = E.C' + E.E$ and $0 = C'.(C'+E) = C'.C' + C'.E$. As by construction C' and intersect in two points (i.e. $C'.E = 2$), it follows that $C'.C' = E.E = -2$. Notice also that, being $\to Q$ a family over a polydisk Q, $\Phi(C'+E) = \Phi$ for any sheaf Φ. The tensor product $\omega_\pi \otimes \mathcal{L}_\pi^{-2}$ trivial off the central fibre and therefore we must have $\omega_\pi \otimes \mathcal{L} = \mathcal{O}(mC' + nE)$ for some integers $, n$. From the relations above, it is easy to compute the degrees

$$d_{C'} := deg\mathcal{O}(mC' + nE)|_{C'} = mC'.C' + nE.C' = -2m + 2n$$

$$d_E := deg\mathcal{O}(mC' + nE)|_E = mC'.E + nE.E = 2m - 2n$$

prove a), notice that $deg\omega_\pi = 2g - 2 = degL_{C'}^2$ yielding $d_{C'} = 0$, that is $m = n$, and $\mathcal{O}(m(C'+E))$ trivial. As for b) the same reasoning leads to $d_{C'} = 2$, $d_E = -2$, i.e. $n = m+1$, and $\mathcal{O}(m(C' + + E) = \mathcal{O}(E)$.

This result generalizes quite nicely what is usually meant by 'plumbing fixture' in the physical erature. In, sticking to the case of a single separating node, we have the following situation. The rmalization of C has two components C_i of genera g_i, $(i = 1, 2)$ with $g_1 + g_2 = g$ and the dualizing eaf of C restricts to $\omega_i(p_i)$, on C_i. As these have odd degree, only b) applies in this case. In particular $(E)|_{C'} = \mathcal{L}^2(p_1 + p_2)$ restricts to C_i to $\mathcal{L}^2(p_i)$ which is isomorphic to $\omega_i(p_i)$. Hence, giving such imit θ-characteristic on C_0 is tantamount to choosing θ-characteristics on the components C_i. We ve then $2^{2g_1}.2^{2g_2} = 2^{2g}$ non-equivalent choices.

A less common picture arises for a single non-separating node, where both cases a) and b) apply. is is to be expected as the genus of C' is $g-1$ and the number of θ-characteristics on it is only a arter of what one would like to have. The correct number is restored on C_0 in the following way. If $|_E$ is trivial, $\mathcal{L}_\pi|_{C'}$ is one of the $2^{2(g-1)}$ square roots of $\omega|_{C'}(p_1 + p_2)$. Notice that these do not come m θ-characteristics on the normalization of C. An extra factor of two is given by the two different ntifications between the stalks on on the points p_i, yielding in total a half of what we need. The t comes in the same way when $\mathcal{L}_\pi|_E$ is $\mathcal{O}(1)$.

Prop. 3.2 tells us that we can get a line bundle as a limit of a family of θ-characteristics by simply wing up the nodes on a family of stable curves. Actually this is not always necessary because, when $|_E$ is trivial, one can safely blow down the exceptional component E, reverting to the previous family. ese are precisely the θ-characteristics which have already a limit as line bundles on families of stable ves. In general, however, one has to deal with families of semi- stable curves. Luckily enough they er the game with extra data, leading to the notion of 'spin-curves' [C] as triples $(C, \mathcal{L}, \phi)$, where is a semistable curve with disjoint rational components E_i, $\mathcal{L}$ is the sheaf of sections of a line ndle of degree $g-1$ on C such that $\mathcal{L}|_{E_i} = \mathcal{O}(1)$, $\phi : \mathcal{L}^2 \to \omega_C$ is a homomorphism vanishing on E_i's. These generalize the one-node case and allow a compactification $\overline{S}_g$ of the moduli space S_g

of θ-characteristics on smooth curves, much alike the Deligne-Mumford compactification of ordina moduli spaces.

There is a natural relation between this compactification and that constructed by Deligne. Inde one gets (reduced) Deligne's deformations by simply blowing down the exceptional lines. As this do not affect the base space of the deformation itself, one gets $\overline{S}_g$ also in this way.

There is no need to describe here more details of this compactification, which can be found in full systematic setting in [C]. It will be enough for us to list the following results;

1) $\overline{S}_g$ has a natural structure of a normal projective variety, $\partial\overline{S}_g = \overline{S}_g - S_g$ is a closed proper analyt subvariety of $\overline{S}_g$, and therefore S_g is an open subvariety.

2) The natural map $\chi : \overline{S}_g \to \overline{S}_g$ given by forgetting spin structures and reverting to stable mod (i.e. blowing down all exceptional components) is finite.

3) Since the parity of a θ-characteristics is invariant under deformations, $\overline{S}_g$ is the disjoint uni $\overline{S}_g = \overline{S}_g^+ \sqcup \overline{S}_g^-$ of the two closed irreducible subvarieties of even and odd spin curves of genus g.

The boundary of $\overline{S_g^{\pm}}$ consists of the following divisors;

- ν_0, made of one-node curves with a square root of the canonical bundle (case a) of Prop. 3.2

- ν_0', consisting of classes of semistable curves with one-node irreducible model and with invertible free sheaf corresponding to case b) of Prop. 3.2,

- ν_i, $i > 0$, parametrizing classes of semistable curves with stable model consisting of two com nents of genus i and $g - i$ and with an $\mathcal{L}$ as in case b) of Prop. 3.2.

Here we denote in the same way the boundary classes of both even and odd spin moduli spac although one should distinguish between e.g. ν_0^+ and ν_0^-. For instance in the even (odd) case, consists of semistables curves with an $\mathcal{L}$ restricting on the two components to both even or odd (even and one odd) θ- characteristics. Also, denoting with δ_i the pull-back to $\overline{S_g^{\pm}}$ of the bound classes of $\overline{S}_g$ consisting of stable curves with components of genera i and $g - i$, it holds [C] δ_0 $\nu_0 + 2\nu_0'$, $\delta_i = 2\nu_i$. To grasp these relations, notice that $\delta = \sum \delta_i$ coincides with the image of nodes of the 'universal curve'. This has precisely one node over ν_0 and two nodes over all the ot boundary components.

A useful tool in controlling the behaviour of determinants of $\overline{\partial}$-operators in the Polyakov boso string was the Grothendieck-Riemann-Roch theorem. We will now briefly see how this can be app in the present situation. As usual we will pretend that there exists the universal curve C over s moduli spaces, in order to avoid technical subtleties which are beyond the scope of this paper. W we are going to say is actually rigorous if one restricts himself to work on the open and dense subvar made of spin curves without automorphisms, or if one deals with the so called 'moduli stack'.

Let then $\pi : C \to \overline{S}_g$ be the 'universal' spin curve of genus g. This comes together with invertible sheaf L_π representing the 'universal' spin structure. On C we have as well the relat structure sheaf O_π and the relative dualizing sheaf ω_π. Recall that, if we have a family of rela $\overline{\partial}$-operators coupled to an invertible sheaf F on C, its determinant $det\overline{\partial}$ is a section of a 'line bun $det\,\pi_! F$ on $\overline{S}_g$ with first Chern class

$$c_1(\pi_! F) = \lambda + \pi_*(\frac{1}{2}c_1(F).c_1(F)) - \pi_*(\frac{1}{2}c_1(F).c_1(\omega_\pi))$$

Here we are following the setup and the notations given, f.i., in appendix A of [H]. We simply no that '.' denotes intersection in homology, π_* is the Gysin homomorphism given by pushing forw homology classes (this operation is the homological counterpart of the fiber integrals which enters 'De Rham' version of the family index theorem) and $c_1(F)$ in homology is the divisor associate the invertible sheaf det F. We also set $\lambda := c_1(\pi_!\omega_\pi)$ for the Hodge class of $\overline{S}_g$. As in [DM], one fi that

$$\pi_*(c_1(\omega_\pi).c_1(\omega_\pi)) = 12\lambda - \delta$$

where $\delta = \sum \delta_i$ is the boundary class.

In fermionic string theory, one is interested in computing Chern classes of integral powers of L. Mumford's formula still applies, yielding the following relation

$$c_1(\pi_!(L_\pi^{2s})) = (6s^2 - 6s + 1)\lambda - \frac{1}{2}(s^2 - s)\nu_0 - (2s^2 - s)\nu'$$

where $\nu' = \nu_0' + \sum \nu_i$ is the boundary class corresponding to semistable spin curves with exceptional components. For instance, the four determinants of $\overline{\partial}$-operators entering the heterotic string theory can be easily seen to be sections of line bundles with Chern classes

$$c_1(\pi_! O_\pi) = \lambda$$

$$c_1(\pi_! \Omega_\pi^{-1}) = 13\lambda - 2\delta$$

$$c_1(\pi_! L_\pi) = -\frac{1}{2}\lambda + \frac{1}{8}\nu_0$$

$$c_1(\pi_! L_\pi^{-1}) = \frac{11}{2}\lambda - \frac{3}{8}\nu_0 - \nu'$$

for a single chiral boson*, for the ghost determinant, for one chiral fermion and for the superghost determinant respectively.

Putting things together, we see that the supersymmetric sector of the heterotic string partition function gives us a contribution of

$$-5c_1(\pi_! O_\pi) + c_1(\Omega_\pi^{-1}) + 5c_1(\pi_! L_\pi) - c_1(\pi_! L_\pi^{-1}) = -\nu_0 - 3\nu'$$

for $g \geq 2$. The case $g = 1$ has to be treated in a separate way; because of the identity $12\lambda = \delta$, one checks that the behaviour at the boundary of the chiral partition function as given by the Grothendieck-Riemann-Roch theorem is in full agreement with the well known explicit computations.

4. Supermoduli space building.

From the discussion outlined in the previous sections, it should be clear that one has a good control of two of the three ingredients needed to define a supermanifold. In fact, what is still lacking, is a complete understanding of how to glue local patches of supermoduli space. In particular, restricting oneself to work with purely even objects is not correct, since a deformation depending trivially on odd parameters has identically vanishing odd super Kodaira-Spencer map, and thus its basis cannot be taken as building block for supermoduli space.

In the following we will restrict ourselves to describe quite informally some of the ingredients entering the construction of the graded analogue of the moduli stack. In practice, we will forget about the existence of automorphism, and pretend that universal deformations $X \to \Delta$ of susy-curves give "coordinate charts" on "supermoduli spaces". Our strategy to get some insight to the geometry of these "spaces" is first to select some very special classes of versal deformations, and then trying to glue their bases requiring that a superconformal isomorphism exists between the families.

First we give concrete examples of versal deformations of a susy-curve. To this purpose we need the following two lemmas

Lemma 4.1 . Let $p \in C_{red}$ be a generic point, then for $n \geq 1$ the connecting homomorphism $\delta: \mathbb{C}^{(n+1)(g-1)} \to H^1(C_{red}, \mathcal{L}^{-n})$ associated to the exact sequence

$$0 \to \mathcal{L}^{-n} \to \mathcal{L}^{-n}((n+1)(g-1)p) \to \mathcal{L}^{-n}((n+1)(g-1)p)/\mathcal{L}^{-n} \to 0$$

is an isomorphism.

---

* Notice that since the sheaf of Kähler differentials Ω_π is not locally free, there is an extra contribution of $\pi_* c_2(\Omega_\pi) = \delta$ to the above version of Mumford's formula applied to ω^{-1}, yielding the correct result of 2δ instead of δ.

Proof. A segment of the long cohomology sequence reads

$$\ldots \to H^0(C^{red}, \mathcal{L}^{-n}(Np)) \to \mathbf{C}^N \to H^1(C_{red}, \mathcal{L}^{-n}) \to H^1(C_{red}, \mathcal{L}^{-n}(Np)) \to \ldots$$

($N = (n+1)(g-1)$) and since the first and the last space have the same dimension, we ne only to prove that one of them vanishes. By Serre duality, this is the same thing as showing th $H^0(C_{red}, \mathcal{L}^{n+2}(-Np)) = 0$, that is that there are no sections of $\mathcal{L}^{n+2}$ vanishing of order $\geq N$ at p. L σ_0 be a local trivializing section of $\mathcal{L}$ around p, and $\sigma_i = f_i(z)\sigma_0$ ($i = 1, \ldots, N$) be the local expressi for a basis of $H^0(C_{red}, \mathcal{L}^{n+2})$. The matrix

$$\begin{pmatrix} f_1(z) & f_1'(z) & \ldots & f_1^{(N-1)} \\ . & . & \ldots & . \\ . & . & \ldots & . \\ f_N(z) & f_N'(z) & \ldots & f_N^{(N-1)} \end{pmatrix}$$

has vanishing determinant whenever one of the f_i's vanishes of order $\geq N$ at p. This cannot be t case for almost all $p \in C_{red}$ because in this case a line of the matrix above would be linear combinati of the others, i.e. we would get a differential equation of order $N-1$ with N linearly independe solutions.

Lemma 4.2 . For a generic point $p \in C_{red}$ and $n \geq 1$, the connecting homomorphism $\delta_p^n : \mathbf{C}$ $H^1(C_{red}, \mathcal{L}^{-n})$ associated to the exact sequence

$$0 \to \mathcal{L}^{-n} \to \mathcal{L}^{-n}(p) \to \mathcal{L}^{-n}(p)/\mathcal{L}^{-n} \to 0$$

is injective. The map $\delta^n : C_{red} \to H^1(C_{red}, \mathcal{L}^{-n})$ given by $p \to \delta_p^n(1)$ is full, i.e. there are $(n+1)(g-$ points p_i such that the classes $\delta_{pi}^n(1)$ form a basis of $H^1(C_{red}, \mathcal{L}^{-n})$.

Proof. The relevant cohomology sequence reads

$$\ldots \to H^0(C_{red}, \mathcal{L}^{-n}(p)) \to \mathbf{C} \to H^1(C_{red}\mathcal{L}^{-n}) \to \ldots$$

Since $deg\mathcal{L}^{-n}(p) = n + 1 - ng$ is negative for $g \geq 1 + 1/n$, injectivity follows at any p for $g \geq 2$ $n > 1$. The same is true in the case $n = 1$ at $g = 2$ because, if $\mathcal{L}$ is even its divisor is not effec and $\mathcal{L}^{-1}(p)$ cannot be trivial. In the odd sector $\mathcal{L}$ has as divisor one of the six Weierstrass po and again $H^0(C_{red}, \mathcal{L}^{-1}(p)) = 0$, provided p is not a Weierstrass point. To show that δ^n is full, enough to notice that if $Im\delta^n$ was contained in a hyperplane in $H^1(C_{red}, \mathcal{L}^{-n})$, then there would an element ϕ of the dual space $H^1(C_{red}, \mathcal{L}^{-n})^\vee = H^0(C_{red}, \mathcal{L}^{n+2})$ such that $< \phi, \delta_p^n(1) >= 0$ fo $p \in C_{red}$. Here $< ., . >$ is Serre duality i.e. $< \phi, \delta_p^n(1) >= res_p\phi.\sigma$ where σ is a representative of δ i.e. a section of $\mathcal{L}^{-n}$ with a first order pole at p. Then this would imply that ϕ itself vanishes, absurdity.

Example 4.3 . A very simple example of a versal deformation of a susy-curve $C = (C_{red}, \mathcal{L})$ ca constructed by concentrating the deformation at a generic point $p \in C_{red}$. Let $\{U_\alpha, z_\alpha, \theta_\alpha\}$ be an a for C and assume $p \in U_0$ with $z_0(p) = 0$. We glue a superdisk with coordinates x_0, ϕ_0 with $C_{red} -$ by means of the map

$$\begin{cases} x_0 = z_0 + \sum_{i=1}^{3g-3} \frac{t_i}{z_0^i} + \theta_0 \sqrt{1 - \sum_{i=1}^{3g-3} \frac{it_i}{z_0^{i+1}}} \quad \sum_{k=1}^{2g-2} \frac{\epsilon_k}{z_0^k} \\ \phi_0 = \sqrt{1 - \sum_{i=1}^{3g-3} \frac{it_i}{z_0^{i+1}} - \sum_{k=1}^{2g-2} \frac{\epsilon_k}{z_0^k} \sum_{k=1}^{2g-2} \frac{k\epsilon_k}{z_0^{k+1}}} \quad \theta_0 + \sum_{k=1}^{2g-2} \frac{\epsilon_k}{z_0^k} \end{cases}$$

with (t_i, ϵ_k) in a small superpolydisk Δ. Now

$$KS_0(\frac{\partial}{\partial t_i}) = [\frac{1}{z_0^i}\frac{\partial}{\partial z_0}] = \rho_p^2(e_i)$$

$$KS_1(\frac{\partial}{\partial \epsilon_k}) = [\frac{1}{z_0^k}\theta_0 \otimes \frac{\partial}{\partial z_0}] = \rho_p^1(e_k)$$

here $\{e_i\}$ and $\{e_k\}$ are standard basis in $\mathbf{C}^{3g-3}$ and $\mathbf{C}^{2g-2}$ respectively. Lemma 4.1 then tells us at KS is an isomorphism and our family is versal.

xample 4.4 . Another class of versal deformations of C can be associated to $5g-5$ generic points p_i. his is closer to what is done in the physical literature (see, e.g. [B]), as it corresponds to considering avitino zero modes as δ-functions on $2g-2$ distinct points. We glue superdisks with coordinates $_i, \phi_i$ with $C - \{p_i\}$ by means of the maps

$$\begin{cases} x_i = z_i + \frac{t_i}{z_i} \\ \phi_i = \sqrt{1 - \frac{t_i}{z_i^2}}\,\theta_i \end{cases}$$

r $i = 1, ..., 3g-3$ and by

$$\begin{cases} x_i = z_i + \theta_i \frac{\epsilon_{i-3g+3}}{z_i} \\ \phi_i = \theta_i + \frac{\epsilon_{i-3g+3}}{z_i} \end{cases}$$

r $i = 3g-2, ..., 5g-5$. Then

$$KS_0(\frac{\partial}{\partial t_i}) = [\frac{1}{z_i}\frac{\partial}{\partial z_i}] = \delta_{pi}^2(1), \qquad i = 1, ..., 3g-3$$

$$KS_1(\frac{\partial}{\partial \epsilon_{i-3g+3}}) = [\frac{1}{z_i}\theta_0 \otimes \frac{\partial}{\partial z_i}] = \delta_{pi}^1(1), \qquad i = 3g-2, ..., 5g-5.$$

gain, Lemma 4.2 tells us this family is modular.

Both these examples yield "local coordinates" (up to automorphisms of the central fibre) on permoduli "space" by

$$\Delta \xrightarrow{\Psi} \hat{S}_g$$

here $\Psi(t,\epsilon)$ =[isomorphism class of $\pi^{-1}(t,\epsilon)$] and we consider on $\Psi(\Delta)$ the sheaf $\Psi_* \mathcal{O}_\Delta$. Whenever o such "charts" overlap, i.e. $X_k \to \Delta_k$ $(k = 1,2)$ are deformations of C_k such that $\Psi_1(\Delta_1) \cap \Psi(\Delta_2) =$ $\neq \emptyset$, then the restrictions X_k' of X_k to $\Psi_k^{-1}(V)$ are isomorphic as families of susy-curves, that is ere are maps g, h making the diagram

$$\begin{array}{ccc} X_1' & \xrightarrow{g} & X_2' \\ \downarrow & & \downarrow \\ \Psi_1^{-1}(V) & \xrightarrow{h} & \Psi_2^{-1}(V) \end{array}$$

mmute. The map h is then the "clutching" function for these two charts on supermoduli space.

To get a closer insight to the structure of such an h, we fix a representative C of a point in V d consider X_k' as deformations of C. This operation will be called "shifting the central fibre" and well defined since, given C, we have isomorphisms i_k of C to the fibres of X_k' over $\Psi_k^{-1}([C])$. We ve also maps $f_k : \Psi_k^{-1}(V) \to \Delta_k'$ inducing X_k' as $f_k^* Y_k$, where $Y_k \to \Delta_k'$ are versal deformations of of the form given in one of the examples above. Then h makes the diagram

$$\begin{array}{ccccccc} & & Y_1 & & \xrightarrow{g'} & & Y_2 \\ & \nearrow & | & & & \nearrow & \\ X_1' & & \xrightarrow{g} & & X_2' & & \downarrow \\ & & \downarrow & & & & \\ & & \Delta_1' & & \xrightarrow{h'} & & \Delta_2' \\ \downarrow & \nearrow & & & \downarrow & \nearrow & \\ \Psi_1^{-1}(V) & & \xrightarrow{h} & & \Psi_2^{-1}(V) & & \end{array}$$

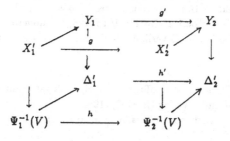

commutative, where (g', h') is an isomorphism of deformations of the form given in the exampl above. So $h = f_2^{-1} h' f_1$.

An easy but tedious computation shows that while the maps f_k can be realized as split ma (i.e. as maps preserving the Z-grading of structure sheaves), the map h' is not even projected. As consequence the "atlases" on $\hat{S}$ given by the examples 4.2 or 4.3 are non-projected. A full proof this fact will be given elsewhere, as we need more machinery which is out of the size of this paper.

Of course this negative result does not give the ultimate answer, as one cannot yet exclude th more sophisticated universal deformations could yield projected or even split "atlases". Indeed at l genus one has independent arguments to show that this is indeed the case.

Proposition 4.6 . Supermoduli space splits in genus $g = 1$.

Proof. On families of elliptic curves one either has no holomorphic 3/2-differentials (in the ca of even θ-characteristics) or, when $\mathcal{L}_\pi \simeq \mathcal{O}_{C_t}$, only one, so that splitness is insured by dimensio reasons.

Proposition 4.7 . Supermoduli space (for *smooth* curves) splits in genus $g = 2$.

Proof. Here, no matter which is the parity of the θ-characteristics , we have that the local model supermoduli $\hat{S}_2$ is the rank two sheaf $\mathcal{E} \equiv R^1\pi_*(\mathcal{L}^{-1})$ over the moduli space of genus 2 spin curves Then (see [R1]) its obstruction to splitness is measured by a single class $\tau_2 \in H^1(S_2, \mathcal{T}S_2 \otimes \wedge^2(\mathcal{E}$ Here we can argue as follows. The natural map $S_2 \to M_2$ obtained by forgetting spin structures finite. But M_2 is an affine variety and hence is Stein. As being a Stein space is a property which preserved in both senses under finite maps, then also S_2 is Stein and so, as $\mathcal{T}S_2 \otimes \wedge^2(\mathcal{E})$ is coher analytic τ_2 vanishes.

5. Concluding remarks

As we have seen, there is still a long way to grasp the global structure of supermoduli spac The main result of this paper is that the simplest choices one can make, and markedly those wh are usually done in the physical literature, yield non projected 'atlases' on supermoduli space. T case of genus 2 is in a way exceptional, as one can show [FR] that one can judiciously arrange thi in such a way that deformations of the form 4.3 give a projected (and hence split) structure to Unfortunately this does not work at higher genera, with the implication, for instance, that the na Berezin integral may not be given any sensible meaning in these complex atlases.

As for the integration problems, one could very well limit oneself to work on the "supermo stack" as we have been morally doing up to now. However, to produce a coarse supermoduli sp we have to face the problem of automorphisms of susy-curves. Besides the automorphisms of underlying spin curves, which make the reduced supermoduli space a complex space indeed, th is the canonical automorphism of the Z_2-graded sheaves involved in the construction which possi make this coarse "space" to be not even a ringed space, but a "superorbifold" (see e.g. [LR]). We that working with deformations may give a clear cut to this problem as well.

Finally, more detailed study of deformations of singular curves should be taken into accoun refine the analysis above and get a compactified version of supermoduli spaces.

Acknowledgements. We are greatly indebted to M.Cornalba for most of the results of sec We are also grateful to M.Martellini and P.Teofilatto for joint work on the problem of superst path integrals, which greatly stimulated our interest on supermoduli spaces. We would like thank L.Alvarez-Gaume, C.Bartocci, M.Bonini, U.Bruzzo, R.Catenacci, E.Gava, C.Gomez, R.Je A.Morozov, P.Nelson, M.Rothstein, E.Verlinde, H.Verlinde for several stimulating discussions.

References

[ACGH] E. Arbarello, M. Cornalba, P. Griffiths, J. Harris, Geometry of algebraic curves Vol I, Gr Math. Wiss. **267**, Springer Verlag (Berlin), (1986).

[B] M. Bershadsky, Super–Riemann surfaces, loop measure etc. ... Nucl. Phys. **B 310**, 79, (19

]] M. Cornalba, Moduli of curves and theta-characteristics. *Preprint* , Universita' di Pavia, (1988).
)] P. Deligne, unpublished letter to Yu.I. Manin, (1987).
)M] P. Deligne, D. Mumford, The irreducibility of the space of curves of given genus. *Publ. Math. I.H.E.S.* **36**, 75 (1969).
)P] E. D'Hoker, D.H. Phong, The geometry of string perturbation theory. *Preprint*, to appear in *Rev. Mod. Phys.* **60** (1988).
'] D. Friedan, Notes on string theory and two dimensional conformal field theory. In *Unified String Theories* , M. Green, D. Gross eds. (1986) World Scientific (Singapore).
'MS] D. Friedan, E. Martinec, S. Shenker, Conformal invariance, supergravity and string theory. Nucl. Phys. **B 271** (1986) 93.
'R] G. Falqui and C. Reina, (in preparation).
;N] S.B.Giddings, P.Nelson, The geometry of super Riemann surfaces *Commun. Math. Phys.* **116**, 607, (1988).
[] R.Hartshorne, Algebraic Geometry, GTM 52 Springer Verlag (Berlin), (1977).
[o] P.S. Howe, Super Weyl transformations in two dimensions. *J. Phys. A: Math. Gen.*, **12**, 393 (1979).
] D.A. Leites, Introduction to the theory of supermanifolds. *Russ. Math. Surveys* **35**, 1 (1980).
R] C. LeBrun, M. Rothstein, Moduli of Super Riemann Surfaces. *Commun. Math. Phys.* **117**, 159 (1988).
[T] M. Martellini, P. Teofilatto, Global structure of the superstring partition function and resolution of the supermoduli measure ambiguity. *Phys. Lett.* **211B** , 293 (1988).
[] Yu.I. Manin, Critical dimensions of the string theories and the dualizing sheaf on the moduli space of (super) curves. *Funct. Anal. Appl.* **20**, 244 (1987).
1] M. Rothstein, Deformations of complex supermanifolds. *Proc. Amer. Math. Soc.* **95**, 255 (1985).
2] M. Rothstein, Integration on noncompact supermanifolds. *Trans. Amer. Math. Soc.* **299**, 387 (1987).
/] A.Yu. Waintrob, Deformations of complex structures on supermanifolds, *Seminar on supermanifolds no 24* D. Leites ed., ISSN 0348-7662, University of Stockholm (1988).

C.I.M.E. Session on "Global Geometry and Mathematical Physics"

List of Participants

T. ACKERMANN, Bergst. 42, 69 Heidelberg

L. ALVAREZ GAUME', CERN, Theoretical Division, CH 1211 Génève 23

V. ANCONA, Istituto Matematico U. Dini, Viale Morgagni 67/A, 50134 Firenze

E. ARBARELLO, Dipartimento di Matematica, Università "La Sapienza", P.le A. Moro 5, 00185 Roma

S. AXELROD, Department of Mathematics, Princeton University, Fine Hall, Princeton, NJ 08544

F. BASTIANELLI, Physics Department, SUNY at Stony Brook, New York 11794

F. BATTAGLIA, Via Scozia 6, 47037 Rimini (FO)

M. BOCHICCHIO, I.T.P., Department of Physics, SUNY at Stony Brooks, NYLI 11794

L. BOURAOUI, 1 rue Montpellier, 91300 Massy

U. BRUZZO, Dipartimento di Matematica, Via L.B. Alberti 4, 16132 Genova

D. CANGEMI, BSP Université, CH-1015 Lausanne

L. CASTELLANI, CERN, Theoretical Division, CH 1211 Génève 23

N. CHAIR, SISSA, Strada Costiera 11, 34014 Trieste

R. CIANCI, Dipartimento di Matematica, Via L.B. Alberti 4, 16132 Genova

E. COLOMBO, Piazza Petrarca 14, 44100 Ferrara

L. DABROWSKI, SISSA, Strada Costiera 11, 34014 Trieste

M. DAMNJANOVIC, Department of Physics and Meteorology, P.O.B. 550, 11001 Beograd

R. D'AURIA, Dipartimento di Fisica Teorica, Università, 35100 Padova

P. DE BARTOLOMEIS, Via Felice Fonatan 7, 50144 Firenze

S. DE LILLO, Dipartimento di Fisica, Università, 84100 Salerno

S. DE MICHELIS, Corso Italia 71, 13039 Trino (VC)

M. DJURDEVIC, Department of PHysics and Meteorology, P.O.B. 550, 11001 Beograd

G. FALQUI, SISSA, Strada Costiera 11, 34014 Trieste

M. FERRARIS, Dipartimento di Matematica, Via Ospedale 72, 09100 Cagliari

T. FLA, IMR, University of Tromso, P.O.Box 953, N-9001 Tromso

M. FRANCAVIGLIA, Istituto di Fisica Matematica, Università di Torino Via Carlo Alberto 10, 10123 Torino

P. FRE', Dipartimento di Fisica Teorica, Via P. Giuria 1, 10125 Torino

O. GARCIA-PRADA, Department of Mathematics, Rice University, P.O. Box 1892, Houston, Texas 77251

. GATTO, Via Serena 41, 10090 Sangano (Torino)

. GERGONDEY, 148 rue de la Louvière, 59800 Lille

. GHERARDELLI, Istituto Matematico U.Dini, Viale Morgagni 67/A, 50134 Firenze

. GIACHETTI, Via Bolognese 39, 50139 Firenze

. GIERES, Inst. f. Theor. Physik, Univ. Bern, Sidlerstr. 5, CH-3012 Bern

. GRAVESEN, IMFUFA, Institute of Mathematics and Physics,
Roskilde University Center Hus 02, Postbox 260, DK-4000 Roskilde

. GRECO, Dipartimento di Matematica del Politecnico,
Corso Duca degli Abruzzi 24, 10129 Torino

.J. HITCHIN, Fellow Street, Saint Catherine College, Oxford OX1 3UJ

. HORVATHY, Dept. de Math. et d'Inform., Metz University, F-57045 Metz Cedex

. LANDI, Ist. Naz. Fis. Nucl., Mostra d'Oltremare, Pad. 19, 80125 Napoli

. LUGO, SISSA, Strada Costiera 11, 34014 Trieste

. LUSANNA, Sez. IMFN di Firenze, Largo E. Fermi 2, 50135 Firenze

. MAGNANO, SISSA, Strada Costiera 11, 34014 Trieste

. MASBAUM, Institut de Mathématique et d'Informatique, Université de Nantes,
44072 Nantes Cedex 03

. MATONE, SISSA, Strada Costiera 11, 34014 Trieste

. MIGLIORINI, Istituto Matematico U. Dini, Viale Morgagni 67/A, 50134 Firenze

MILLER, Math. Inst. II, 7500 Karlsruhe

MULLER, Im Neuenheimer Feld 683, 69 Heidelberg

MUSSO, Via Squarcialupo 19/A int. 8, 00100 Roma

NANNICINI, Istituto di Matematica Applicata, Via S.Marta 3, 50139 Firenze

NICOLODI, Via della Chiesa 6, 38060 Aldeno (Trento)

NYKANEN, NORDITA, Blegdamsvej 17, D-2100 Copenhagen

PAOLETTI, Collegio Ghislieri, Piazza Ghislieri 5, 27100 Pavia

PARRINELLO, Dipartimento di Fisica, Università "La Sapienza",
P.le Aldo Moro 2, 00185 Roma

PEKONEN, Centre de Mathématiques, Ecole Polytechnique, F-01128 Palaiseau Cedex

-E. PFISTER, Department of Mathematics, EPT, CH-1015 Lausanne

PIERZCHALSKI, Institute of Mathematics, Lodz University,
ul. Banacha 22, 90-238 Lodz

M. RASSIAS, 4 Zagoras Street, Paradissos, Amaroussion, 15125 Athens

REINA, SISSA, Strada Costiera 11, 34014 Trieste

RICKMAN, Department of Mathematics, University of Helsingki,
Hallituskatu 15, 00100 Helsinki

ROGORA, Via Firenze 13, 20025 Legnano (Milano)

M. ROTHSTEIN, Department of Mathematics, University of New York at Stony Brook, NYLI 11794

B. RUNGE, Universitat Mannheim, Fak. fur Math. und Inf., D-6800 Mannheim 1

J. RUSSO, SISSA-ISAS, Strada Costiera 11, 34014 Trieste

M. SCHLICHENMAIER, Universitat Mannheim, Fak. fur Math. und Inf., Lst.II, Seminargebaude A5, D-6800 Mannheim 1

W. K. SEILER, Universitat Mannheim, Fak. fur Math. und Inf., Postfach 10 34 62, D-6800 Mannheim 1

M. SEPPALA, University of Helsinki, Department of Mathematics, Hallituskatu 15, SF-00100 Helsinki

S. SHNIDER, Department of Mathematics, Ben Gurion University, POB 653, Israel

D.-J. SMIT, Institute for Theoretical Physics, University of Utrecht, P.O.Box 80.006, 3508 TA Utrecht

I. SOLS, Universidad Complutense de Madrid, Facultad de Ciencias, Departamento de Algebra y Fundamentos, Madrid

R. STORA, CERN, Theoretical Division, CH 1211 Génève 23

E. STRAUME, Institute of Physics and Mathematical Sciences, University of Tromso, N-9001 Tromso

V. TAPIA, SISSA, Strada Costiera 11, 34014 Trieste

F. TRICERRI, Istituto Matematico U. Dini, Viale Morgagni 67/A, 50134 Firenze

G. WEILL, Faculté des Sciences, Université de Tours, Dept. de Mathématiques, Parc De Grandmont, 37200 Tours